D1502565

ABOUT ISLAND PRESS

Island Press, a nonprofit organization, publishes, markets, and distributes the most advanced thinking on the conservation of our natural resources—books about soil, land, water, forests, wildlife, and hazardous and toxic wastes. These books are practical tools used by public officials, business and industry leaders, natural resource managers, and concerned citizens working to solve both local and global resource problems.

Founded in 1978, Island Press reorganized in 1984 to meet the increasing demand for substantive books on all resource-related issues. Island Press publishes and distributes under its own imprint and offers these services to other nonprofit organizations.

Support for Island Press is provided by Apple Computers, Inc., Mary Reynolds Babcock Foundation, Geraldine R. Dodge Foundation, The Educational Foundation of America, The Charles Engelhard Foundation, The Ford Foundation, Glen Eagles Foundation, The George Gund Foundation, William and Flora Hewlett Foundation, The Joyce Foundation, The J. M. Kaplan Fund, The John D. and Catherine T. MacArthur Foundation, The Andrew W. Mellon Foundation, The Joyce Mertz-Gilmore Foundation, The New-Land Foundation, The Jessie Smith Noyes Foundation, The J. N. Pew, Jr. Charitable Trust, Alida Rockefeller, The Rockefeller Brothers Fund, The Florence and John Schumann Foundation, The Tides Foundation, and individual donors.

ABOUT NATURAL RESOURCES
DEFENSE COUNCIL

Natural Resources Defense Council is a nonprofit membership organization with more than 130,000 members and contributors nationwide, and with offices in Washington, D.C., San Francisco, Los Angeles, and Honolulu. Since 1970, NRDC scientists and lawyers have been working to protect America's natural resources and to improve the quality of the human environment. NRDC pursues these goals through shaping government action at the federal, state, and local levels and through extensive public education and outreach. NRDC's major accomplishments have been in the fields of public health (including air and water pollution and pesticide safety), resource conservation (including coastal protection and agricultural conservation), energy conservation, and the international environment (including the proliferation of nuclear weapons and global warming).

In addition to its national and international activities, NRDC has long held a special interest on the urban environment in general and New York City in particular. Over the last two decades, NRDC has fought successfully for city improvements in public transportation, expansion of urban parklands, reductions in motor vehicle and industrial air pollution, and increased programs for recycling. New initiatives to decrease sewage discharges into our rivers and bays, to limit unchecked development projects, and to document the worse environmental problems in the city's low-income communities are now underway. NRDC believes that its New York City program, and a similar effort underway in its new Los Angeles office, can serve as a model for similar efforts in urban areas around the nation.

The New York Environment Book

NATURAL RESOURCES DEFENSE COUNCIL

*Eric A. Goldstein
and Mark A. Izeman*

ISLAND PRESS

Washington, D.C. □ *Covelo, California*

© 1990 Natural Resources Defense Council

All rights reserved. No part of this book may be reproduced in any form or by any means without permission in writing from the publisher: Island Press, Suite 300, 1718 Connecticut Avenue NW, Washington, D.C. 20009

Library of Congress Cataloging-in-Publication Data

Goldstein, Eric.
The New York environment book / Natural Resources Defense Council : Eric Goldstein and Mark Izeman.
p.   cm.
ISBN 1-55963-019-1 (alk. paper). — ISBN 1-55963-018-3 (pbk. : alk. paper)
1. Environmental protection—New York.   2. Environmental policy— New York.   I. Izeman, Mark.   II. Natural Resources Defense Council.   III. Title.
TD171.3.N5G64   1990
363.7′009747′1—dc20   90-4024
CIP

Printed on recycled, acid-free paper

Manufactured in the United States of America

10   9   8   7   6   5   4   3   2   1

To our parents

and to John H. Adams,
who, as executive director of the
Natural Resources Defense Council for 20 years,
has inspired a generation of lawyers and scientists
to fight to protect the nation's environment
and the health of its citizens,
including the 7.2 million residents of the
City of New York

Acknowledgments

We could never, ever have completed this book alone. Our long list of thank-yous starts appropriately enough with our colleagues here at the Natural Resources Defense Council. Our very own Urban Environment Program staff were extremely supportive throughout the three-year research and writing effort. When in the fall of 1989 we were most exhausted and thought that the manuscript might never be finished, in came Samuel A. Hartwell, a first-rate research associate, who graciously helped us in so many ways to cross the finish line. Very able research assistance in earlier phases of our work was provided by Louis Weinberg, Melissa Paly, Andrew Kass, and James Murphy, all of whom have since gone on to respectable careers in their own right.

Mitchell Bernard and Katherine Kennedy provided helpful edits and, more importantly, kept the Urban Program sailing smoothly as we were tied up with this publication. And Wallace Boulton, who typed and retyped the entire manuscript and kept the same crazy hours we did, performed with talent above and beyond the call of duty.

Many of our colleagues throughout NRDC offered insights and suggestions to sections of the book dealing with issues for which they are legitimately known as national experts. They include David Doniger, Allen Hershkowitz, Robert F. Kennedy, Jr., Jessica Landman, Mary Nichols, Frederica Perera, Nina Sankovitch, Debbie Sheiman, Lisa Speer, and Jacqueline Warren. Peter Borrelli and Paul Allen cheerfully guided us through the publishing process, for which we are most appreciative.

As much as any single person, Nevin Cohen, staff member to New York City councilman Sheldon S. Leffler, made this book possible. He read every page of every chapter we wrote, offering new ideas and constructive criticisms along the way. His wealth of information about New York City environmental issues and his unselfish commitment of time to this project improved the book immeasurably. We can't thank him enough.

We also called upon other outside experts, many of whom devoted considerable time to reviewing early drafts of this publication. Among those to whom we owe a special thanks are Robert Alpern, Albert Appleton, Marjorie J. Clarke, Adam Cohen, Vincent Coluccio, Ralph Hallo, Cara Lee, Dr. Steven B. Markowitz, Granville Sewell, Carol Steinsapir, and Philip Weinberg.

There are plenty of people working for federal, state, and city government agencies who care deeply about New York City's environment. More than a handful have been extremely helpful to us throughout the course of our research. Deserving special thanks are Ivan Braun, Jack Lauber, Jim Meyer, Benjamin Miller, Alan Mytelka, James Repace, Berry Shore, and Glenn Rubinstein. Many others assisted us with data-gathering and offered confidential appraisals of their agency's activities. For obvious reasons, we shall keep their names to ourselves, but express our thanks to them collectively here.

We can't say enough about our editors at Island Press. Karen Berger, Nancy Seidule, and Robin Barker, with whom we have jousted on more than one occasion, came through for us and published this book in record time. Everyone at Island Press has earned our deep appreciation.

We are grateful too, to our former colleagues Ross Sandler and David Schoenbrod, who had the foresight to create NRDC's Urban

Environment Program more than 15 years ago. Their idea of assembling a team of lawyers and scientists who would work exclusively to protect environmental quality in New York City continues to flourish.

We owe an immense debt of thanks to the foundations and philanthropic supporters who have stood by us over the years. Without each and every one of them, publication of this book would simply have not occurred: Louis and Anne Abrons Foundation, Inc.; The Achelis Foundation; The Vincent Astor Foundation; Booth Ferris Foundation; The Clark Foundation; Robert Sterling Clark Foundation, Inc.; The Aaron Diamond Foundation; The Educational Foundation of America; The J. M. Kaplan Fund, Inc.; Dextra Baldwin McGonagle Foundation, Inc.; Morgan Guaranty Trust Company of New York Charitable Trust; Henry and Lucy Moses Fund, Inc.; The New York Community Trust; The New York Times Company Foundation, Inc.; Jessie Smith Noyes Foundation; The Prospect Hill Foundation; Charles H. Revson Foundation, Inc.; the members of STAND; the family of Lawrence Wein; and Robert W. Wilson.

Finally, we thank our family and friends, who have given us their loving support while patiently awaiting the completion of this book. All of us now realize that it is easier to read a book than it is to write one.

Contents

Foreword   **xv**

Introduction   **xvii**

Part One: Solid Waste   **1**

Dwindling landfills, controversial incinerators, and infant recycling and waste reduction programs—say hello to New York City's most pressing environmental headache.

Part Two: Waterways and the Coast   **45**

Even as 14 treatment plants are reducing sewage discharges, there are plenty of reasons to worry about New York City's rivers, bays, and coastline.

Part Three: Air Pollution   **85**

Despite some improvements over the last two decades, air quality in New York City is nothing to brag about; motor vehicles and other small sources are still making big problems for your lungs.

Part Four: Drinking Water   129

You probably don't think much about New York City's single most valuable capital asset, but its high-quality drinking water supply may now face bigger threats than at any time this century.

Part Five: Toxics   165

Lead and asbestos in the home present well-publicized risks; yet what could be the most serious toxic threat for many New Yorkers may surprise you.

Part Six: Epilogue   211

A treasure chest of ideas for how citizens can take things into their own hands and begin cleaning up New York's environment.

Index   253

About the Authors   265

Foreword

I ask myself, "Why has a book about New York City this basic never existed before?" It is as though only now somebody got around to wondering and then describing how the insides of an animal's body worked. Only now do we have this physiology of the enormous creature millions of us inhabit, for better or for worse, as though we were tiny, unthinking parasites. If the creature sickens, then we sicken, too. So why haven't even schoolchildren, never mind grown-ups, been told what its guts need and do?

I think I may have answered my own question by using the word "guts," which implies smelly gases, bad breath, urine, and appalling quantities of excrement. Who wouldn't rather hear about loftier matters, such as art and politics and high finance or whatever? And most public persons, including journalists and politicians, know better than to offend their audiences with serious, informed discussions of the disgusting bodily functions of the drinking, eating, sweating, belching, wind-breaking, hugely excreting host animal, older than the city of Leningrad, incidentally, on which our health depends.

Be that as it may, and in the immortal words at the end of Arthur Miller's New York tragedy, *Death of a Salesman*, "Attention must be paid."

Yours truly,

Kurt Vonnegut

Introduction

When we started poking around in our library at the Natural Resources Defense Council (NRDC) several years ago, looking for a book about environmental problems in New York City, we were somewhat surprised that we could not find any. After all, it wasn't as if there was nothing to write about. The city's air quality had a well-earned reputation as among the nation's worst. Hundreds of millions of gallons of raw sewage were pouring into city waterways every day, even as government agencies sought to complete the country's largest urban network of sewage treatment plants. And local officials, pointing to shrinking capacity at city landfills, were already declaring a garbage disposal crisis.

Still, about all we could turn up were folders of newspaper and magazine articles on such issues, and a series of thoughtful booklets put out in the mid-1970s by the Mayor's Council on the Environment of New York City. We decided right then and there to try to fill the gap.

This is the first book to define the debate on environmental issues in New York City in the 1990s. It divides the environmental problems New Yorkers are facing into five categories—solid waste disposal, waterways and the coast, air quality, drinking water, and toxics—and organizes its discussion around these groupings. For each issue, it summarizes the problem's importance, marshals available data on environmental trends, offers a capsule of applicable laws, and highlights government progress or lack thereof in turning the situation around.

Why zero in on these five issues? Clean air and clean waterways are widely acknowledged as among the most fundamental environmental issues of the last 20 years. Drinking water has similarly been a long-standing focal point of environmental concern. Although toxics issues have been added to the environmental agenda more recently, the significance of lead and asbestos pollution and of occupational health issues in New York City prompted us to include a full discussion of such topics in this volume. Finally, we discuss solid waste both because it has become one of the city's most visible environmental problems and because New Yorkers will probably have to be more involved in solutions to the waste disposal dilemma than any other environmental challenge the city now faces.

Of course, this book does not cover every item on New York City's environmental platter. We hardly touch, for example, on the city's parks, where, fortunately, talented citywide groups like the Parks Council and local watchdogs like the Alley Pond Environmental Center are keeping a protective eye on things. Similarly, we have only skimmed the issue of public transportation, an area to which NRDC previously devoted considerable energy; at least this matter is now receiving greater public scrutiny, as evidenced by the activities of the New York Public Interest Research Group citizen-based Straphanger's Campaign and by stepped-up media attention. And the need to draw the line somewhere led us to omit the issue of noise pollution, a decision we regret whenever we are caught in a subway station as a screeching express train rounds the bend.

During our three-year research effort, we conducted more than 100 personal and telephone interviews. We completed a detailed review of government reports. And we filed several dozen freedom of information requests for unpublished data. But in too many areas, we were struck by the paucity of government statistics, monitoring, and record-keeping. We are not the first to warn that sound environmental decision-making requires a more complete and accessible information base. We will probably not be the last.

Although this book focuses on New York City, we suspect that there are more than a few similarities between our city's problems and those of, say, Los Angeles or Chicago. New York is not alone in its struggle to cope with mounting trash, to clean tainted waterways, to tame air pollution, and to protect its populace from toxins in the home and in the workplace. Nothing would make us happier than to see this book serve as a springboard for analyses by citizen groups in urban areas across the nation.

Everyone, it seems, loves lists. "Which are the most impressive environmental success stories?" our friends and colleagues would inquire when they heard we were writing this book. "Which do you rank as the city's worst environmental problems?" we've been asked more than once. For list-lovers and for speedsters who can satisfy themselves with only the most fleeting glimpse at an issue, we offer NRDC's subjective index to the best and worst of New York City's environment.

The Best:

- The city's upstate reservoir system, which provides, on average, roughly 1.5 billion gallons of water every day and remains the envy of drinking water providers around the country;
- The 14-sewage-treatment-plant system, which, despite troubling problems, has significantly cut back on discharges of raw sewage into New York City's rivers and bays;
- The still-suffering-but-on-the-way-up bus and subway network, without which many of its roughly 2.5 million daily riders would turn to automobiles, choking the city in pollution and congestion; and
- The federal program virtually eliminating lead from gasoline, an important success in the continuing struggle to eradicate childhood lead poisoning in this city.

The Worst:

- Slow progress in recycling trash, and reducing the amount of wastes generated in the first place;
- Continuing discharges of raw sewage from nearly 500 outfalls in the city's ancient system of combined storm and sanitary sewer pipes;
- Insidious toxic pollution of our waterways from industrial, commercial and residential discharges, including approximately 7,000 pounds a day of heavy metals;
- Continuing growth in drinking water consumption from city reservoirs (30 percent higher today than in 1960) and the city's inability thus far to mount an effective, long-term water conservation program here;
- Inappropriate development that threatens city neighborhoods and our upstate watersheds;
- Unhealthy levels of air pollutants spewing from automobiles, buses, and trucks, whose travel throughout the region has shot up by more than 20 percent since 1970;
- Toxic air emissions from approximately 2,200 apartment house, scores of hospital, and three antiquated municipal incinerators, most located in densely populated neighborhoods;
- Preventable exposures to leaded paint (most seriously affecting children in low-income neighborhoods in Brooklyn and the Bronx) and to asbestos (present in roughly two-thirds of all city buildings); and
- Workplace exposures to environmental toxics, probably the most overlooked toxic risk for many New Yorkers today.

We close this book with an important epilogue. In it, we spell out the first steps New Yorkers can take right now to improve and protect public health and our urban environment. Meanwhile, we invite you to hang on for a ride through New York City's environmental battlegrounds.

PART 1

Solid Waste

To many public officials have been asking the wrong questions about solid waste in New York. They have focused almost exclusively on how to replace dwindling landfill space to bail themselves out of the city's waste disposal quagmire. But New York City does not have a single garbage crisis. It has three of them.

The first, to be sure, is a capacity crunch. Every day during 1989, more than 14,000 tons of trash—close to 75 percent of the total disposed of in the city—were being dumped at a sole landfill (the Fresh Kills burial grounds on Staten Island). This has left the city too dependent on a single site and vulnerable to the capacity limits of that facility. Sanitation officials justly call the situation desperate.

Seeking relief, the city has proposed dotting the boroughs with at least five large garbage-burning incinerators. But this approach ignores the second crisis—a logistical one. Unless New Yorkers recycle more and reduce the amount of trash they generate, total garbage loads in this city will soon outstrip all disposal capabilities. The arithmetic is simple. Even if the city's proposed incineration network of nearly 12,000 tons a day of capacity is actually constructed, almost 40 percent of the trash would still have to be handled by other means. And that doesn't count the 30 percent of

3

incinerated waste that remains as toxic ash residue (and, of course, the adverse air quality and ash disposal impacts of widespread garbage burning).

There is, in addition, a third crisis—the environmental and health risks posed by former and still-operating landfills and by existing apartment house, hospital, and municipal incinerators. Cleanup of many landfills closed since the 1970s has been stalled despite the inclusion of these dumps on the state's list of highly contaminated disposal sites. And government scientists quietly concede that uncontrolled apartment house and hospital incinerators, along with their poorly performing municipal cousins, have created localized air pollution problems of particular severity.

Some environmental problems can be worked out between government regulators and affected industries themselves. Solid waste is not one of them. New Yorkers have direct control over their trash through what they purchase and what they throw away. The amount of waste generated, the toxicity of materials discarded, and the level of recycling and waste reduction are all affected by citizen decisions. Thus, New Yorkers could have a greater impact in solving the solid waste problem than in perhaps any other environmental issue facing the city today.

A SOLID WASTE SNAPSHOT

You could think of it as Garbage City. A staggering 19,000 tons of waste are disposed of in the Big Apple every day. (For years, the number hovered around 27,000 tons a day. The recent fall-off is largely a result of higher dumping fees at city landfills, which has prompted commercial carters to dispose of their trash truckloads outside city limits.) On the average, every New Yorker discards about five pounds of trash every 24 hours. This is roughly twice as much as that produced by the typical Japanese, Swede, German, Spaniard, Swiss, or Norwegian.[1]

Until recently, only half of the waste disposed of in the city came from private homes and apartments. But since the city boosted landfill dumping fees, commercial trash, construction rubble, street and bulk debris, and other associated wastes have slid from roughly 50 to less than 30 percent of the total.[2] In comparison, infectious medical wastes represent roughly one-third of one per-

FIGURE 1.1: WHERE THE CITY'S TRASH GOES

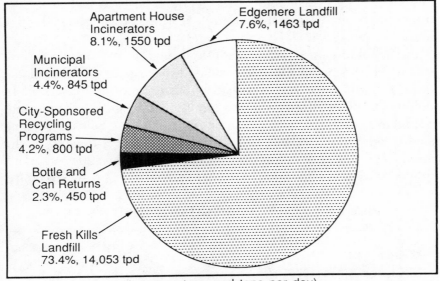

Apartment House
Incinerators
8.1%, 1550 tpd

Edgemere Landfill
7.6%, 1463 tpd

Municipal
Incinerators
4.4%, 845 tpd

City-Sponsored
Recycling
Programs
4.2%, 800 tpd

Bottle and
Can Returns
2.3%, 450 tpd

Fresh Kills
Landfill
73.4%, 14,053 tpd

(in percentage and tons per day)

NRDC (October 1989)

cent of the overall waste load. (In this chapter, our references to solid waste loads are usually described by weight, not volume, since most figures are reported using this measure.)

In New York City today, waste disposal means landfilling almost by definition. Staten Island's infamous Fresh Kills landfill crams in close to 75 percent of the city's daily refuse. With the exception of what is brought to the much smaller Edgemere landfill in Queens, most of the city's remaining trash is burned in aging incinerators. Recycling brings up the rear. A mere 6.5 percent of the city's waste stream is now being set aside through state and city recycling programs, according to the latest Sanitation Department data, and more than a third of that amount comes from bottle and can returns (see figure 1.1.)

LANDFILLS

Background

For a time, it seemed like the perfect solution. What were thought of by some as unsightly swamps and unsanitary wetlands could be filled in to create new streets and building sites. For much of this

century, landfills offered New York officials a place to dispose of the
bulk of the city's trash and expand its boundaries as well. In 1934,
the number of city-operated landfills scattered throughout the five
boroughs peaked at 89.[3]

But by mid-century, the landfilling engine had begun to lose
steam. Many smaller fills had reached their capacity, and available
space for siting new facilities had become increasingly scarce. In
the 1960s and 1970s, as concern mounted over garbage dumping,
public officials formulated the first environmental rules governing
landfill operations. These factors helped trigger a decline in the
number of New York dump sites. And during the last 20 years, not a
single new landfill has opened its gates anywhere in New York City.

FIGURE 1.2: NEW YORK CITY LANDFILL SITES

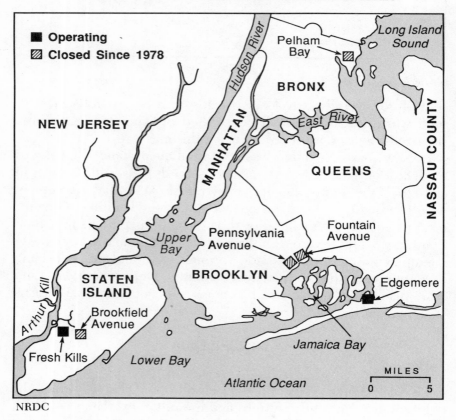

NRDC

Nobody, however, should kiss landfills goodbye. Although sanita-tion officials warn that the end of landfilling at existing locations is now in sight, dates on which the city's two active dump sites will actually close remain uncertain. And with proposed garbage-burning incinerators likely to generate 3,600 tons a day or more of ash residue, the city cannot be expected to give up its toehold in the existing landfills without a fight. At the same time, state environ-mental officials and citizen groups are unlikely to let the cleanup of old landfill sites slip entirely off the Sanitation Department's agenda.

Landfills: Environmental and Public Health Impacts

No one really knows the full story of the environmental and public health risks posed by city landfills. Government officials have only examined the potential adverse impacts at the two existing landfill sites and at a handful of the most recently closed facilities. Even at these locations, studies have been limited. Here are the highlights.

SURFACE WATER POLLUTION

The experts call it leachate. What it consists of is rainwater that has filtered through a landfill site, often collecting toxic contaminants along the way. The leachate passes into surrounding waters either as direct runoff from refuse mounds or through underground seep-age. By anyone's standards, there's a lot of it. The Fresh Kills landfill alone may be releasing as much as 2 million gallons a day of leach-ate. Additionally, another million gallons are excreted daily by the Edgemere and several recently closed city landfills.[4]

It is no surprise, then, that contamination of adjacent surface waters is the single most significant adverse impact from city land-fills. Not much data have been gathered. But monitoring in 1982 found high levels of mercury in waters surrounding four city land-fills in Brooklyn, Queens, and the Bronx. Mercury readings were well in excess of state and federal water quality standards. Samples taken in 1983 from Jamaica Bay, near two of these facilities—the Fountain Avenue and Pennsylvania Avenue landfills—found levels of lead and nickel that were nearly 100 times greater than state standards. PCBs and other toxic chemicals have also been detected in these environmentally sensitive waters. In fact, national park officials who now oversee this area have singled out leachate from these two old dumps as a major threat to the ecosystem of Jamaica Bay.[5]

GROUNDWATER POLLUTION

Maybe New York City doesn't have to worry about landfills polluting its groundwater? Maybe the city is off the hook since it now gets virtually all of its drinking water from upstate reservoirs, not underground water supplies? And maybe officials are correct in assuming that pollution from existing landfills will not get into the city's most important underground water tables? Maybe, but maybe not.

Contaminating groundwater should never be taken lightly. And tests have shown that outer portions of the Brooklyn-Queens aquifer underneath several landfill sites are already polluted with cadmium and lead at levels that exceed state drinking water standards. Benzene, trichloroethylene, and other organic compounds have also been discovered.[6] The city does not presently draw water from this edge of the aquifer. But once pollutants taint underground waters, they are exceedingly difficult to remove. The long-term risk from contamination of this resource, a conceivable source of drinking water in future decades, should not be quickly discounted.

AIR POLLUTION

The extent of the air quality problem posed by city landfills has not been determined. Methane, produced from the decomposition of refuse, is probably the single largest component of landfill gases. While methane does not appear to present a localized public health threat, this landfill gas is one of a handful of substances contributing to the gradual warming of the earth's lower atmosphere known as the "greenhouse effect." Volatile organic compounds, which contribute to ozone smog and include trace elements of carcinogens such as benzene, also emanate from city landfills, although city officials suggest that such emissions are insignificant and represent no threat to the public.

Perhaps the most comprehensive public health assessment of

landfill air emissions in New York City was a 1982 study of nearly 2,000 individuals living near Staten Island's inactive Brookfield Avenue landfill. It found a statistically higher rate of respiratory problems (sore throats, coughs, and upper-respiratory-tract infections), but no significant differences in total doctor visits or hospitalizations. Toxic air pollutants have also been monitored directly above one other landfill (Fountain Avenue), making occupational exposures at the sites themselves perhaps the most compelling concern.[7]

The Law: Landfills

In the 1970s, state and federal lawmakers sought to bring the age-old practice of landfilling into the modern era. Rules were drafted to govern the safe operation of active landfills and insure the sound management of closed sites.

As in many other areas of solid waste regulation, the state, not the federal government, was first to act. In 1973, the legislature authorized the Department of Environmental Conservation to develop rules governing landfill operations. Four years later, the department adopted specific criteria that all landfill operators were required to meet. As now amended, the rules govern everything from landfill security to water pollution monitoring to procedures for facility closure. In theory, state-issued permits spell out the site-specific details.[8]

The ambiguously named 1976 Resource Conservation and Recovery Act (RCRA) opened solid waste landfills to federal regulation. In its most recent amendments, the statute requires state permit programs to include landfill standards at least as stringent as those adopted by U.S. Environmental Protection Agency (EPA). The new provisions also direct EPA to toughen requirements for solid waste landfills receiving toxic wastes produced by household and small-quantity generators.[9]

Report Card on the City's Landfills

It's a good thing there is no official report card on New York City landfill operations. If there were, you'd find more F's than A's. Neither of the city's two active landfill sites, Fresh Kills and Edgemere, currently is meeting state rules that, among other things, prohibit the flow of leachate into surface and ground waters. Nor have four other recently shut down New York City landfills complied with regulations mandating proper closure and long-term pollution monitoring. In addition, five of the six sites are listed officially as state hazardous waste dumps. [10]

A wide gap exists between legislative objectives and day-to-day realities in this area. Why? First, city officials originally sited many landfills in ecologically fragile wetland areas. While placing landfills on the city's fringes created new real estate and eased landfill siting, it made sealing off leaking dumps an expensive proposition. Second, the practicalities of disposing of more than 19,000 tons of refuse every day has made it more difficult for state officials to bring and pursue aggressively enforcement actions against the city's active landfills. Finally, city control of inactive landfill sites has created a deadlock between state and local officials over who should foot the bill for landfill cleanup.

FRESH KILLS

This is one distinction New York officials would gladly do without. The city's Fresh Kills landfill is reportedly the world's largest. It stretches over western Staten Island for nearly 3,000 acres and it receives over 14,000 tons a day of residential and commercial refuse. It is home to roughly about a third of the total waste now being landfilled in New York State.

The world's biggest landfill, however, has no state operating permit. Uncontrolled pollutants streaming into surrounding waters is one reason. For example, black plumes of leachate containing cyanide and other toxins have recently been observed snaking their way from Fresh Kills into the nearby waters of the Arthur Kill. [11]

City and state officials have for years been arm wrestling over how to bring Fresh Kills into compliance with state landfill rules. Under a 1985 agreement with the state, city officials were required, among other things, to improve on-site trash handling, spruce up landscaping, and control leachate flows. A series of federal court orders dating back to 1983 have directed the city to prevent Fresh Kills refuse from washing onto New Jersey shores. [12] And in the

Wide World Photos

END OF THE LINE. *Last stop for much of New York's trash is the Fresh Kills landfill on Staten Island. The 3,000-acre dump, reportedly the world's largest, crams in up to 75 percent of the city's total garbage output. Most trash arrives on city-owned barges filled up at dockside transfer stations located throughout the city. Early next century, Fresh Kills will be filled to capacity, big trouble for New Yorkers unless alternative waste disposal arrangements (like waste reduction and recycling) are made soon.*

most recent flurry of activity, the state filed a new administrative enforcement action in late 1989 to secure an accelerated Fresh Kills clean up and closure. Despite some improvements at the be-leaguered dump, however, Fresh Kills remains open largely for one reason—there is simply no other major city facility to handle so much garbage.

When will Fresh Kills cease operations? Nobody really knows. In 1985, sanitation officials projected that the facility would be filled to capacity by the year 2000. More recent estimates are that the city may be able to shoehorn garbage into Fresh Kills until 2020, or beyond.[13] (The new forecasts are based, in part, upon a recent fall-off in the amount of commercial waste being brought to Fresh Kills. This decline has paralleled higher landfill dumping fees imposed on private haulers.) Solid waste planning in New York has never been divorced from local politics. It is likely, then, that political considerations, as much as capacity limitations and regulatory requirements, will determine when the last garbage barge docks at the Fresh Kills landfill.

EDGEMERE

In New York, the Edgemere landfill has often been seen as small potatoes. But many city managers across the country would consider this facility a big-time operation. Making its home on the Rockaway Peninsula in Queens, the landfill is 173 acres in size—more than 17 times larger than Shea Stadium. Every day in 1989 nearly 1,500 tons of household trash were trucked into this dump site.

State environmental officials have been seeking for years to close the Edgemere landfill. Their main concern has been environmental risks posed by toxic wastes disposed of illegally at the site. Since 1983, more than 3,000 drums of potentially toxic material have been unearthed there. This and other evidence have prompted the state to rank the facility as a top priority for hazardous waste cleanup.[14] Meanwhile, Edgemere continues to violate state regulations governing proper landfill operations. It remains open without a valid state permit.

Like Fresh Kills, Edgemere's closure date is up in the air. As part of a 1987 agreement with state officials, the city consented to phase out landfilling at this site by 1991. But New York City decisionmakers may be eyeing Edgemere for possible use as an incinerator ash disposal site to handle residues if new garbage-burning plants come on line. And the loosely worded state-city agreement would appear to leave this option open.[15]

FOUNTAIN AVENUE

Brooklyn's Fountain Avenue landfill is the most recent municipal site to close. This 297-acre dump is located on the edge of Jamaica Bay, south of the Belt Parkway and the Starrett City residential complex. From its opening in 1961 to its closure in 1985, the landfill received construction and demolition debris, residential trash, asbestos and ash residue from municipal incinerators. In its final year of operation, Fountain Avenue took in an average 8,200 tons of trash every day, second only to Fresh Kills.

Like other recently shut down city landfills, Fountain Avenue has never satisfied state environmental rules for safely closing and securing the site. Ironically, the property was transferred to the National Park Service (for inclusion in Gateway National Recreation Area) before on-site cleanup activities were completed. Today, the Fountain Avenue landfill remains on the state's list of hazardous waste sites warranting priority attention. And Park Service scientists have reported detecting PCBs on the property. Although city officials have signed an agreement to mop up any solid or hazardous waste problems at the landfill, they maintain that the present situation poses no major threat. [16]

PENNSYLVANIA AVENUE

The Pennsylvania Avenue landfill, which lies next to the Fountain Avenue site along Jamaica Bay, has been closed for a decade. The book on its environmental problems, however, remains open. This 110-acre facility began operations in 1956 to handle regular residential and commercial wastes, but later received primarily sewage sludge, construction and demolition debris. It was accepting between 1,000 to 2,000 tons a day when landfilling ended there in 1979. Now, it is also part of the Gateway Recreation Area.

Environmental contaminants linger at the Pennsylvania Avenue landfill. Government investigators have uncovered 6 to 12 million gallons or more of waste oil inside the landfill; they fear that it could seep into adjacent surface waters. Already, oily leachate containing PCBs and heavy metals has been found along the landfill site, oozing into Jamaica Bay. [17]

MEDICAL WASTE

Has extensive television and other media attention blown up New York's medical waste disposal problem out of proportion?

If you look only at the small amount of medical debris that has actually washed ashore and consider that such events have not yet had widespread health consequences, you'd be tempted to say yes. But consider the absurdity of summer beach closings. Or the related environmental problems of poorly controlled hospital incinerators. Bring in factors like these and the picture comes into sharper focus. Media attention in this complicated area has not always hit dead center. Yet there is little doubt that the broad issue of medical waste disposal deserves a continuing place in the spotlight.

How big is the medical waste problem? A 1988 New York State Health Department study estimated that 73 tons of "infectious waste" is generated every day by hospitals, nursing homes, clinical laboratories, and diagnostic and treatment centers in New York City. But this tabulation does not include wastes from smaller sources such as doctors' and dentists' offices. And by counting only "infectious wastes," the Health Department identified just a portion of all medical wastes whose improper disposal may pose health or environmental problems. The amount of medical trash is growing—thanks in part to the AIDS crisis and the stepped-up use of disposable products in patient care. Still, even if the total amount of medical waste is three times the state Health Department projections, it is still comparatively small—equaling about 1 percent of the city's total solid waste stream.

To many New Yorkers, it is those medical wastes that have washed ashore on the region's beaches that matter most. Officials estimate that a total of 4,200 pieces of medical waste turned up on city beaches from July to September of 1988. This is roughly equivalent to perhaps a dozen grocery bags of material. Medical products, however, made up only an estimated 1 to 10 percent of all debris that washed ashore that summer. In part as a result of these washups, many New York City and Long Island beaches were closed on 16 separate days. Staten Island beaches were the hardest hit; four beaches (South, Midland, Great Kills, and Miller Field) were out of commission for stretches of two to eight weeks at a time during the 1988 season.[22]

Public reactions to waste

William E. Sauro, New York Times Pictures

TIP OF THE ICEBERG. *Needles washing up on New York beaches grabbed headlines in recent summers, but the issue of medical waste disposal is far more complex. The amount of medical waste generated in hospitals, doctors' offices, nursing homes, and laboratories is growing, in part because of increased reliance on disposables. Waste reduction, a shutdown of inefficient incinerators at individual hospitals, and the construction of a state-of-the-art regional medical waste incinerator will likely be part of the solution.*

washups have been fueled by fear of AIDS and hepatitis, much of it unfounded. The AIDS virus is fragile, and scientists believe that its ability to survive outside the body for more than several hours is close to zero. Hepatitis, a more resilient but usually less serious virus, is more of a concern; to date, however, widespread cases of hepatitis from medical waste washups have not been reported.

This is not to say that the public outrage is unwarranted. Even the small risk of contracting disease or of being injured by medical sharps on a Sunday beach outing is widely considered unacceptable. And the closing of ocean beaches, often justified under these circumstances, is among the most frustrating impacts of environmental pollution in New York.

The manhunt for medical waste polluters has turned up some unexpected suspects. A state Department of Environmental Conservation report has concluded that sewer system overflows, and city landfill and garbage barge operations, rather than illegal dumping, were primarily responsible for beach washups in the summer of 1988. City officials, understandably defensive, are unwilling to downplay the contribution of midnight dumpers. But they acknowledge that the city's sewer and garbage systems are not completely innocent.

The public outcry over beach pollution has whipped up further interest in the incineration of medical wastes—a disposal method favored by many hospitals and other medical waste generators. Most state and city officials have been quick to endorse the incineration alternative. According to the State Health Department's 1988 report (which is hardly definitive), scores of facilities in New York City are apparently burning over 20 tons a day of infectious wastes alone. And new or proposed incinerators are already moving forward at a half dozen or more hospitals around the city.[23]

But incineration is not a panacea for the medical community. Existing hospital incinerators suffer from the same deficiencies as existing New York City apartment

A 900-foot, sausagelike, floating boom has been added to Jamaica Bay's landscape. Its purpose: to contain escaping waste oils from the Pennsylvania Avenue landfill. It is perhaps the most visible example of Sanitation Department action to combat hazardous waste pollution at a city landfill facility. But it is only a stopgap measure (and apparently ineffective, at that). City engineers have been unable to solve the problem at its root—they cannot recover large waste oil deposits that lie within the landfill itself. As with other inactive dump sites, state and city officials have drawn up a consent agreement to tackle contamination problems at the landfill.[18] The problems, however, are likely to fester for years.

house trash-burning facilities: poorly controlled burning; absence of sophisticated antipollution equipment; reliance on untrained operators; and low stack heights, which maximize public exposure to airborne emissions, especially in densely populated neighborhoods. Significantly, the proportion of plastics, including polyvinyl chlorides or PVCs (the rigid molded plastic used for such items as disposable syringes), may be as much as four times higher in hospital waste than in everyday household trash. Some experts believe that the high percentage of these plastics boosts the amount of pollutants, including dioxin, that are emitted from hospital incinerators.[24] Under new state rules that tighten emissions standards, most, if not all, existing on-site hospital incinerators will have to stop burning by 1992.

While it is the antiquated equipment that is the most troubling aspect of older hospital incinerators, for new facilities the concerns are different. Skyrocketing medical costs are leading some hospitals to install less than state-of-the-art incinerators, and to build incinerators with capacities large enough to handle not only infectious wastes but regular hospital trash. This approach yields unnecessary air emissions and undercuts the incentive for hospital waste reduction. Look for regional incineration with state-of-the-art pollution controls and a shutdown of inefficient incinerators at individual hospitals to join waste reduction as pillars on which medical waste disposal policies may be built in the 1990s.

BROOKFIELD

There aren't many people who live across the street from a city landfill. Residents along a strip of Staten Island's Great Kills neighborhood are the exception. Abutting their community lies the 272-acre Brookfield landfill, just east of Fresh Kills on the western portion of Staten Island. The facility, closed since 1980, reportedly contains millions of gallons of industrial waste in addition to household trash that was lawfully disposed of there. As described above, a 1983 city Health Department study, completed in response to complaints about odors and fears concerning health risks from nearby residents, found a greater incidence of respiratory problems but no statistically significant increase in serious illnesses. In 1982–83, the Sanitation Department laid a clay cap over a 41-acre sector of the landfill to minimize odors and surface and ground water pollution. The landfill does not yet comply with state environmental rules for permanent shutdown. [19]

PELHAM BAY

There's more than marshland and scenic greenery in Pelham Bay Park. This northern Bronx landmark is also home to the recently closed Pelham Bay landfill, now an 81-acre parcel of barren land that reaches 18 stories at its highest point. Opened in 1963, the landfill took in an average of 2,600 tons of garbage per day until dumping ceased in 1978. It is suspected that significant quantities of hazardous waste were illegally buried there. And the site, like other recently closed city landfills, is listed in one of the state's most urgent categories for hazardous waste cleanup. A 1988 Health Department investigation into reports of childhood leukemia in the vicinity of the landfill found no elevated incidence of the disease. But an overall analysis of the landfill's health and environmental impacts has not been completed by city or state officials. The city, which under a consent order was to have properly shut down the dump by December 1989, began monitoring for toxics at the site in the fall of that year. [20]

OTHER CLOSED LANDFILLS

Most of the city's older inactive landfills have not been the subject of comprehensive analysis. Some may present localized problems. For example, a 1986 city Health Department study of the old South Shore landfill in southwestern Queens (which received ash from a nearby municipal incinerator for years) found lead in soil at concentrations two to ten times those associated with elevated lead levels in children. And at the site of the former Idlewild landfill near Kennedy Airport, PCBs were detected. Groundwater samples there also revealed elevated levels of lead, chromium, and other heavy metals. Development plans are underway at both locations and government officials recommended that safety measures be taken to minimize public health and occupational risks as development at the two sites moved forward.[21]

RECYCLING AND WASTE REDUCTION

Introduction

We'd like to report big gains in the area of New York City recycling. Sorry. Despite a recent burst of legislative activity, recycling is not yet playing a major role in solving the city's solid waste crisis. Today, city recycling programs are recovering roughly a 6.5 percent sliver of the trash disposed of in New York City. And except for deposits on bottles and cans, most New Yorkers are only now beginning to come in contact with government-sponsored recycling programs, years after the city formally committed to move into high gear in this area.

Does the city's Sanitation Department have it in for recycling? This seems unlikely. Even city officials would agree that recycling conserves natural resources, saves energy, and could reduce the amount of materials that would otherwise have to be buried or burned. They also acknowledge that recycling could be economically competitive with landfilling and incineration. But sanitation officials have long devoted the bulk of their attention and limited resources to advancing incineration. And while they ultimately supported the law that requires mandatory recycling, they do not seem convinced that recycling and waste reduction can become the dominant weapon in solving the city's disposal crisis. This threatens to become a self-fulfilling prophecy.

Recycling: The Law

Despite long-term grassroots interest, recycling has only captured the attention of government lawmakers in the last decade. The first statewide push came in 1980. In that year, the legislature directed the Department of Environmental Conservation to adopt an annual solid waste management plan. State legislators fired another round

NEW YORK STATE'S SOLID WASTE MANAGEMENT PLAN

Seven years in the making, New York State's long-awaited Solid Waste Management Plan was finally released by the Department of Environmental Conservation in 1987. The plan formalizes, at least in theory, the state's commitment to a hierarchy that places recycling and waste reduction ahead of incineration and landfilling. It establishes an ambitious ten-year goal of recycling and reducing by 50 percent the trash generated in New York State. And it outlines in broad terms possible strategies for achieving these objectives. But the plan does not include enforceable measures to reach its recycling or waste reduction targets. To be sure, DEC regulations require any new or

expanded incinerator, landfill, or recycling facility to "demonstrate consistency" with the Solid Waste Management Plan. But actual progress in achieving the 50 percent goal will be contingent on further action at both state and local levels.

Enter New York City, which is now preparing a comprehensive garbage disposal blueprint for the first time. Under a 1988 state law, all municipalities must draft long-term solid waste plans to reduce waste generated and spur recycling. Without such plans, the state can hold up permits for new garbage-burning incinerators. For municipal solid waste strategists, the 1990s should be a busy decade.

in 1982, when, with strong support from New York City officials, they passed the bottle bill, which requires a five-cent deposit on all beer and soda containers sold within the state. In the wake of mounting trash problems, hopes were high in 1988 for far-reaching statutory relief. But the most the legislature and the governor could muster was a compromise bill calling on municipalities to enact, by 1992, their own laws for the separation of recyclable materials from household trash.[25]

At the local level, a 1989 city law may finally provide a shot of adrenaline. The new legislation mandates that all New Yorkers separate recyclables from their household and commercial trash. And it establishes a 25 percent five-year recycling goal, which the

Sanitation Department is obliged to attain. The bill was spear-headed by councilmembers Sheldon Leffler and Ruth Messinger (now Manhattan borough president), among others, with early support from former comptroller Harrison Goldin. Additional provisions in the comprehensive statute require composting of leaves and yard wastes, and the establishment of a city program to recover household batteries and tires (in the event no federal or state programs covering these two materials are put in place by late 1990). The bill also contains a soft command that city officials begin long-term planning for marketing recyclables. And it tacks on a directive that may stimulate city agency purchases of products made from recycled materials.[26]

Recycling: Current Status

The little town of East Hampton, Long Island, could teach New York City a big lesson. Aggressive efforts there by local residents have demonstrated the potential power of recycling. A ten-week, 1987 pilot program, organized by Barry Commoner's Center for the Biology of Natural Systems at Queens College, showed that 84 percent of trash (not including bulky items, yard waste, and hazardous materials) collected from 100 participating households could be potentially recycled.[27]

Nobody expects all New York City neighborhoods to mirror the recycling capabilities of a small Long Island suburb. But that doesn't explain the sluggishness of city recycling efforts to date. By late 1989, recycling programs were recovering only about 1,250 tons per day according to city figures. (This does not include private sector recycling, on which little hard data are available.)[28] The 1,250 tons equal about 6.5 percent of the Big Apple's daily refuse load. And more than a third of these gains was attributable to beverage container returns under the state "bottle bill."

We hope that things will look different in 1994. By that date, under the city's new mandatory law, recycling levels should hit 25 percent. The Department of Sanitation has already begun to step up its action on recycling. It has boosted its program staff from roughly 20 to 75. This number is expected to climb to 87 in 1990.

TABLE 1.1 RECYCLING IN NEW YORK CITY (October 1989)*

Program	Tons per Day
Beverage Containers (Bottle Bill)	450**
Commercial Recycling	
Wood chips	240
Glass	25
Curbside Recycling	177
Bulk Recycling	
Appliances, furniture, etc.	100
Dirt (lot cleaning)	68
Asphalt	73
Apartment House Recycling	
City program	54
Environmental Action Coalition	3
R2B2 "Buy-Back" Center (excluding beverage containers)	27
Office Paper Recycling (New York City Council on the Environment)	14
Municipal Agency Office Paper Recycling	14
Volunteer drop-off recycling centers	1
TOTAL	1,246

New York City Department of Sanitation; Recoverable Resources/Boro Bronx 2000, Inc.; Environmental Action Coalition.

 * Does not include unknown tonnage of materials recycled in New York City by private firms not connected with government-affiliated programs.

 ** This number reflects those beverage containers being redeemed for deposit; actual recycling numbers are lower.

And the city's fiscal year 1990 budget for recycling programs is $43 million, a more than twofold increase over 1989.

But translating even the modest five-year, 25 percent recycling mandate into reality will take aggressive city action. The results are not yet assured. And how the city will meet the state's more ambitious 50 percent recycling and waste reduction target by 1997 remains at this point a mystery.

Here's where things stand today on recycling's front lines:

BOTTLE BILL

Some New Yorkers consider the bottle bill to be a real pain in the neck. Apartment dwellers and owners of small groceries bemoan the lack of space to store bottles and cans. Shoppers too often run into problems with reluctant merchants when it comes to cashing in their empty cans and bottles. And retailers sometimes have to wrestle with beverage distributors to secure regular pickups of containers they have redeemed for consumers.

But that's hardly the full story. The bottle bill program is the single largest element in the city's present recycling campaign. Every day, as much as 450 tons of beer and soda containers are returned to supermarkets and other outlets in the city's five boroughs. This has cut down on litter, particularly along highways, parks, and beaches. And recycling bottles and cans helps preserve scarce landfill space.[29]

The city could squeeze even more out of the bottle bill. Return rates for bottles and cans in many upstate regions are hovering around 90 percent. But in the city, levels have apparently not climbed above 60 to 70 percent.[30] (Some watchdogs suggest that even this figure is inflated.) Increased enforcement, additional redemption centers, and more aggressive public education could boost returns by hundreds of tons a day. Sanitation officials are counting on this to happen.

One final thought. High return rates do not necessarily equal high recycling levels. In some cases, beverage containers are simply disposed of as ordinary trash after collection by retailers or distributors. This is less true for aluminum cans and most glass bottles, which enjoy relatively profitable resale markets. Plastic bottles, however, have been a problem. In 1985, roughly two-thirds of all plastic bottles returned in New York State were landfilled. More recent assessments suggest that perhaps as much as 50 percent of plastic beverage containers are now being recycled. This is still far below the estimated 80 percent and 84 percent statewide recycling rates of glass and aluminum cans, respectively. Plastic bottle manufacturers are racing to improve the situation. But as of today not all beverage containers are created equal.[31]

RESIDENTIAL RECYCLING

New York City's residential recycling program has not yet reached its critical mass. The centerpiece of the existing campaign has been the collection of newspapers, bottles, and cans in lower-density neighborhoods. Under this program, residents have been encouraged to separate these recyclables and place them in front of their homes each week. As of October 1989, projects in all five boroughs were collecting an average of 140 tons of newsprint and 36 tons per day of bottles and cans.

Curbside recycling from high-rise apartment buildings is just now hitting stride. The private Environmental Action Coalition, which has been touting recycling for nearly two decades, is now under contract with the city to recycle newspapers and other materials at over 140 apartment buildings in four boroughs. A second apartment house program, launched by the Sanitation Department itself, has expanded this approach to more than 1,570 additional buildings. As of October 1989, these two programs were recapturing an average of 50 tons a day of newsprint and 6 tons of bottles and cans. Curbside and apartment house recycling are slated to expand citywide under the 1989 mandatory recycling statute.

What would happen if city residents were encouraged to recycle not just a handful of materials, but also every conceivable item suitable for such treatment in their household trash? A heterogeneous Manhattan neighborhood is expected to become the testing ground for this so-called "intensive recycling" concept over the next few years. This demonstration project, supported by David Dinkins when he was Manhattan borough president, might just turn out to be the most important experiment in the solid waste arena during the early 1990s.[32]

COMMERCIAL WASTE

Most people are surprised to learn that the city's daily business and commercial waste load could rival trash from homes and apartments. In 1988, roughly 30 percent of trash going to city facilities came from businesses and commercial firms, according to Sanita-

City of New York Dept. of Sanitation

GARBAGE TRUCKS OF TOMORROW. *Trash pickups around the city are chang-ing these days, as the Sanitation Department begins implementation of a mandatory recycling program. New trucks, like the one pictured above, will be collecting separately sorted newspapers, metals, glass, and other recyclable materials, placed out for curbside pickup. The new city recy-cling law mandates that the Sanitation Department achieve a 25 percent recycling rate within five years, but leaves unanswered how the city will attain New York State's 50 percent waste reduction and recycling goal by 1997.*

tion Department figures. (This percentage fell off in 1989, in re-sponse to city increases in landfill "tipping fees.") Additional amounts of this trash—nobody knows how much—is hauled from the city. Thus, recycling of commercial waste could be a big gun in the city's solid waste arsenal. But like much information on private carting, present levels of commercial recycling are held close to the vest.

The mandatory recycling law recognizes the importance of min-ing commercial wastes. It requires the Sanitation Department to designate recyclable materials that represent at least half of the commercial waste stream, and directs businesses (or their private carters) to separate out these materials for recycling. But while the bill articulates an ambitious goal in this area, it is hardly precise in its outline of how this objective is to be accomplished.

OFFICE PAPER RECYCLING

New York City is a white-collar town. And that could be good news for recycling. Paper makes up roughly 85 to 95 percent of all office building trash, by weight. Much of this is computer paper, white ledger, and tab or index cards, which are readily recyclable due to their high quality and to relatively stable market prices. Roughly 750 tons of such first-class office paper are generated in New York City every day, city data reveal.

The city has launched two forays into office paper recycling. The larger one is administered by the Council on the Environment of New York City; it has arranged for the collection of 14 tons a day of high grade and other office paper from the United Nations, Columbia University, Consolidated Edison, the *New York Times*, and over 130 other commercial users. In a separate venture, as many as 80,000 employees are participating in the municipal agency office paper recycling program. It is netting, officials say, roughly 14 tons of paper a day from more than 120 agencies, institutions, and nonprofit groups.

Under the new city law, office paper recycling will likely mushroom, as New York's white-collar industries are pressed to join the recycling crusade.

RECYCLING CENTERS

Depending on whom you ask, the city has either aggressively encouraged or passively observed the development of local recycling centers. In either case, the model program can be found in the Bronx, going under the space-age-sounding name of "R2B2." This facility, privately operated by Bronx 2000, a local development corporation, allows residents to exchange cans, bottles, newspapers, and other materials for cold cash. In 1989, New Yorkers were hauling an average of 27 tons of material a day to this recycling hub, not counting beer and soda containers. The city has also chipped in financial support for voluntary drop-off centers, including the Village Green Recycling Team (Greenwich Village) and the People's Environmental Program (Upper East Side), which collect somewhere around one ton a day of newspapers and other recyclables.

Count on the recycling center infrastructure to multiply under the city's new mandatory program.

MARKETS

It's easy to overlook the importance of markets in the race to recycle. One's first reaction is to focus on how much recyclable material can be collected. But finding markets for bottles, cans, newspapers, and other recyclable items is what could ultimately determine the outer limits of recycling's potential in New York City. Until now, city officials have not had to confront the issue of large-scale marketing head on. But under the new city law, the Sanitation Department must begin long-term market planning. To get the ball rolling, city consultants are figuring out for the first time how much of what is actually in the city's trash. The next step for sanitation officials must be to convince recycling entrepreneurs that the city can deliver a consistent, high-quality supply of recycled materials.

Whether the city's marketing efforts succeed or fail can only be measured by comparing the cost of recycling to other disposal alternatives. This does not mean that the Sanitation Department should abandon efforts to obtain maximum dollars for recyclables. But even markets that throw off only a small income per ton of trash collected look good when the alternative is landfilling or incineration, both of which cost city taxpayers big dollars now, with further increases sure to follow.

PROCUREMENT

New York officials are only now learning lesson number one in the how-to course for developing recycling markets: use the city's purchasing power. The City Council thought the issue was important

WHY PICK ON PLASTICS?

Card-carrying members of the plastics lobby are up in arms these days. They have been incited by growing legislative interest in banning certain plastic products on environmental grounds. In Suffolk County, New York, lawmakers have prohibited the distribution of some nonbiodegradable packaging like plastic grocery bags, as well as food containers and utensils composed of certain plastics. Other jurisdictions are contemplating similar actions. Pending legislation in the New York City Council would discourage the use of disposable utensils and nonrecyclable packaging materials, many of which are plastic.

Are government officials unfairly singling out plastics in frustration over escalating garbage problems? Plastics manufacturers would have you think so. "A helter-skelter rush to anoint villains," scream slick Mobil Corporation ads on the op-ed page of the *New York Times*. And with similar concerns, the Society of the Plastics Industry and related trade organizations jump into court to challenge the new Suffolk County law. [35]

To be sure, part of the motivation for taking on plastics is symbolic. Plastics are among the most visible examples of our throwaway society. And legislative efforts affecting just plastic products can only provide a small measure of relief. (In fact, some of the momentum to restrict plastics was generated not by the waste disposal crisis, but by concerns over the use of ozone-depleting chlorofluorocarbons in the manufacture of polystyrene, or Styrofoam, products.)

But the attack on plastics has been fired from well-buttressed positions. For one thing, plastics have fueled the explosive growth in packaging, which now accounts for fully one-third of the total garbage disposed of in New York State. Plastics are currently the fastest-growing segment of the packaging industry. Although plastics make up only 7 percent by weight of the solid waste stream, they account for as much as 30 percent of our trash by volume. And it's the lightweight plastics that made the disposable revolution possible. Throw-away pens, cigarette lighters, razors, and many other disposables are now piled on store shelves that once stocked only reusables.

To make matters worse, most plastics are difficult to recycle. Products composed of different plastic layers (or combined plastic and nonplastic materials) face technological hurdles to recycling and reuse. And collection, transportation, and reprocessing costs have made the recycling of some plastics uneconomical. (This may help explain why plastic beverage containers returned under New York's bottle bill program are recycled at a lower rate than aluminum cans and most glass bottles.)[36]

Recognizing these vulnerabilities, the plastics industry has drawn up a two-pronged counteroffensive. First, they have argued that it makes sense to incinerate plastics in solid waste garbage-burning plants. But despite their high energy content, plastics, when burned, are thought to be a major source of dioxins and their combustion, surprisingly, unleashes toxic metals as well.[37] (Lead, cadmium, and nickel, for example, are used as color pigments and stabilizers in many plastic products.)

Some proponents of plastics have also unveiled a new weapon—biodegradability. By altering the chemical composition of plastics, manufacturers have come up with ways to shorten the time it takes for these products to break down in the environment. This may be good news for controlling roadside litter and protecting birds and marine life from the dangers of strangulation or ingestion. But as a strategy for easing the solid waste crisis, its bark is bigger than its bite. Biodegradable plastics may complicate the recycling of other plastics by introducing less stable components. They also do nothing to reduce the risks from burning. And since they may not decompose quickly once buried, they are unlikely to make more than a marginal difference in reducing the demand for landfill space.

enough in 1987 to authorize spending up to 10 percent more for city purchases of paper products made from recyclables. But perhaps reflecting the low priority that procurement of recyclables has received, city officials have entered into only a handful of contracts for the purchase of such materials. And in 1989 testimony before the City Council, the Department of General Services sounded less than enthralled with its new duties (under the mandatory recycling law) to speed the purchase of recycled paper and other goods, although it has recently attempted to beef up its staff to comply with the statute.[33]

Waste Reduction: A Postscript

It's more important to the environment than other solid waste control strategies. It consumes no natural resources. And it has no pollution-producing side effects. Even recycling can't make such absolute claims. We are talking here about waste reduction— cutting back on the amount of trash generated by New Yorkers in the first place.

You'd think everyone would be jumping at this idea. But waste reduction initiatives have stalled at the start. Despite considerable discussion and apparently widespread support for the concept, no major program to reduce the amount of waste generated has made it into law. One of the closest jabs so far has come from upstate assemblyman Maurice Hinchey, who in 1988 and 1989 introduced legislation that would provide tax incentives to reduce overpackaging of nonfood products sold within the state.

New Yorkers discard nearly double the amount of trash per person than do residents in parts of Western Europe or in Japan. (Visit European food markets and you'll notice shoppers packing groceries not in plastic or paper bags, but in their own fabric sacks.) Much of the waste New Yorkers produce can easily be trimmed. Package-related materials, for example, now make up roughly one-third of the state's total solid waste load. And the amount of packaging wastes has doubled over the past 30 years and is expected to grow another 25 percent by 1997.[34] Waste reduction remains the wild card in New York City's solid waste future.

INCINERATION

Few issues have seemed to divide the environmental community as much as garbage incineration. Many activists, often neighborhood-based, have given an unambiguous thumbs-down to new garbage-burning facilities in their communities. Several national environmental organizations, on the other hand, have been less absolutist in their opposition to incineration. Not surprisingly, the sponsors of garbage-burning projects and other incinerator proponents have sought to exploit these differences.

But in reality the gulf in New York City between local citizen activists and national environmental advocates is not nearly as large as it might first appear. As noted above, there is a growing consensus that recycling and waste reduction must become the cornerstones of trash disposal policy in coming years. On that, New York's environmental community is in complete agreement. Moreover, neither the city nor the state has yet implemented a model regulatory program to insure that incinerator operations are protective of public health to the maximum degree possible. The environmental community agrees on that point as well. And between its existing, antiquated apartment house and hospital incinerators and its poorly operating municipal burners, the city has yet to compile an incineration track record that inspires much confidence. Here, too, environmentalists agree.

Burning garbage is a risky proposition. And many New Yorkers are justifiably concerned about the air pollution impacts, ash residue disposal problems, and economic burdens that come part and parcel with widespread garbage incineration. If city and state officials do not resolve these public concerns by regulating effectively in the air quality and ash contamination arenas and by proving through their actions that recycling and waste reduction head up their waste disposal priorities, it is likely that proposals for new garbage-burning plants will continue to face stiff opposition in New York City. In this analysis as well, the environmental community speaks with one voice.

Background

Some people talk about incineration as if the concept of trash burning just arrived on the scene. Not so. Actually, New York City was home to the nation's first refuse incinerator, a rudimentary contraption built on Governors Island in the early 1880s. Garbage burning here took off in the 1930s, following a U.S. Supreme Court prohibition on ocean dumping of municipal refuse. And a 1951 city law actually required all new, large apartment houses to contain on-site incinerators. By the 1960s, more than one-third of the city's trash was being burned in roughly 17,000 apartment house incinerators and 22 larger municipal plants scattered throughout the five boroughs.[38]

Garbage Burning Today

Concerned that one day the city will be burning garbage in your neighborhood, using an incinerator without adequate control technology and discharging dangerous pollutants into the air? That day is already here if you live in the Greenpoint or Gravesend sections of Brooklyn, or Maspeth, Queens.

In those three communities, antiquated municipal facilities are burning city trash using technology that is decades behind the state of the art. To be sure, the three incinerators have been operating at less than full capacity in recent years, as engineers tinker with their old-fashioned pollution control equipment. But even when the modifications are complete, these incinerators will not shine like new. In late 1989, the city's three antediluvian municipal facilities were burning 845 tons of trash a day, city data show. This figure could climb to 2,750 tons a day when the patched-up incinerators are fully back on line.

Existing incineration in New York City is not confined to the three municipal dinosaurs. Many hundreds of smaller on-site incinerators are still burning trash in New York's older apartment houses. Fortunately, the number of these incinerators has been dropping. From 17,000, the total is now approximately 2,200; nobody seems to have a precise count. These apartment house incinerators are extremely inefficient. Of the roughly 1,550 to 1,700 tons

Wheelabrator Technologies Inc.

COMING SOON TO A NEIGHBORHOOD NEAR YOU? *Pictured above is West-chester County's garbage incinerator, similar in design to the trash-burning plant proposed for the Brooklyn Navy Yard. The city now operates three antiquated municipal incinerators (two in Brooklyn, one in Queens). Sanitation officials hope to construct five large new trash burners throughout the city (one in each borough), in what would become the highest concentration of garbage-burning plants in the nation.*

per day of refuse that city officials say they burn, nearly half remains as ash residue.[39] It is unceremoniously shipped off to local landfills, with ordinary municipal trash. Fortunately, as is discussed later in this section, the days of incinerators in apartment houses are numbered.

Garbage Burning Tomorrow

Expect the first New York City municipal incinerator constructed in more than 35 years to rise along the East River, at the site of the Brooklyn Navy Yard. But don't make plans for the ribbon-cutting ceremony just yet. Nearly a decade ago, the city picked the Navy Yard as the spot for its first "resource recovery" plant, a euphemism for an incinerator that generates electricity from heat produced during burning. The Navy Yard's proximity to a Con Ed facility (which could purchase the energy generated) and its waterfront access (which could ease garbage drop-off and ash shipments), among other reasons, apparently outweighed concerns about adjacent land uses (which include not only industrial development, but active residential communities).

This siting decision has withstood considerable neighborhood opposition. And the proposed incinerator has undergone lengthy environmental and citizen advisory committee reviews. But the city has not yet secured the official go-ahead to begin construction. In 1988, state Environmental Conservation commissioner Thomas C. Jorling ruled that no permit could issue under state law until sanitation officials had come up with an acceptable plan for safe disposal of incinerator ash and until the city had adopted a comprehensive recycling program. By late 1989, the city had addressed the recycling concerns but had not, the commissioner determined, come up with an acceptable plan for ash disposal. Many observers expect that the city will ultimately satisfy the state objections, but few venture to say when.

As presently designed, the Navy Yard incinerator, with four separate furnaces and a 500-foot stack towering 40 stories over the north Brooklyn landscape, would burn 3,000 tons of trash a day. It would handle 15 percent of the total daily trash now disposed of in the city. And it would generate 935 tons a day of ash residue.[40] Measured by these yardsticks, the $400 million Brooklyn Navy Yard facility would be the largest incinerator in New York State, and one of the largest in the world.

Sanitation kingpins aren't talking much about their long-term citywide incineration plans these days. Who could blame them? Local opposition to the Brooklyn Navy Yard incinerator is still running high. City officials have their hands full just trying to get the first plant on line. One battle at a time, they figure.

But if the Brooklyn Navy Yard plant gets going, you will most likely see movement in the other boroughs. Waiting in the wings are plans for four large new incinerators—Wards Island, Manhattan (2,000 tons a day); Maspeth, Queens (2,000 tons a day); Barretto Point, Bronx (2,000 tons a day); and Arthur Kill, Staten Island (3,000 tons a day). If constructed, these four incinerators and the Brooklyn Navy Yard would receive nearly 12,000 tons of garbage every day, over 60 percent of the total trash disposed of by the city. Since only about 12 percent of city refuse is now incinerated, this ambitious construction program would bring about a dramatic shift

FIGURE 1.3: NEW YORK CITY'S MUNICIPAL INCINERATORS (EXISTING AND PROPOSED)

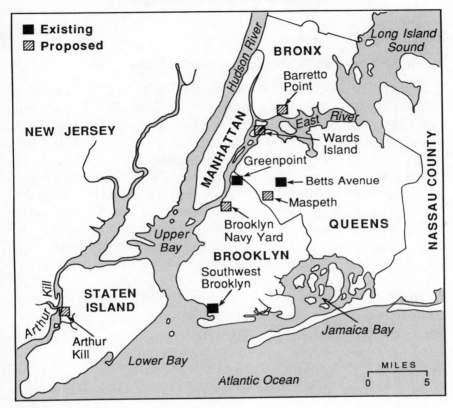

NRDC

in disposal practices. In a matter of years, incineration would replace landfilling as New York's dominant method of waste disposal.

What's more, the number of new garbage-burning incinerators whose plumes might one day be visible on the New York horizon could climb still higher. In 1984, the Sanitation Department reported that the Navy Yard incinerator would be the first of eight resource recovery facilities that would ultimately be fired up in the Big Apple. (The three additional candidate locations identified were Gansevoort Street in Manhattan's West Village; Brooklyn's South Shore; and College Point, Queens.) As noted above, the city already operates three existing municipal incinerators. If sanitation officials have their way, New York City may end the 1990s with the highest concentration of garbage-burning incinerators in the nation.

Trash Burning: Health and Environmental Risks

A heated debate continues to rage on the health and environmental impacts of garbage burning in New York. Sanitation Department officials, in concert with industry spokespersons, maintain that modern, well-run, state-of-the-art incinerators are safe, and present only insignificant risks to public health. Without these incinerators, they maintain, the city will simply be buried in its own garbage. Environmental advocates, not surprisingly, disagree. They point to air emissions and unresolved ash disposal issues posed by incinerators. They are skeptical of the performance promises of incinerator technocrats and worried about the cumulative impacts from a concentration of these facilities throughout the five boroughs.

The issues here are complex and important to the city's future. They deserve a hard look.

AIR POLLUTION

Scores of air contaminants are discharged in varying concentrations as by-products of combustion. The burning of garbage is no exception. Three broad categories of incinerator emissions have aroused the concern of environmental scientists.

One is particulate matter. Particulates may exist as solids or liquid droplets. They range in size from visible black soot to infinitesimal particles as small as 1/250 thousandth of an inch in size. These are sinister pollutants. The smaller, fine particles can penetrate the body's natural defense mechanisms and lodge permanently in the deepest, most sensitive parts of the lung. Lead and such other metals as arsenic, cadmium, and mercury, which may escape during incineration, are frequently carried mainly on such particles. These heavy metals are associated with a wide variety of adverse health impacts ranging from childhood learning disabilities and adult hypertension (lead) to nervous system damage and dangers to the fetus (mercury).[41]

Also wafting from incinerator stacks are acidic gases. Sulfur dioxide and hydrogen chloride are two such contaminants. Their emissions are related to levels of sulfur and chlorine in the waste stream. These corrosive pollutants can aggravate symptoms of heart and lung disease. And along with nitrogen dioxides (another incinerator contaminant), they are the key ingredients of acid rain.

Dioxins are the most infamous of incinerator pollutants. Remember Times Beach, Missouri? The application of dioxin-tainted oil to local roads there led to a public health alert and the 1983 federal buy-out of all homes and businesses in that now desolate community. Nobody is really sure how dioxins (and related furan compounds) are formed during incineration. But poor combustion of certain plastics, papers, and other materials, which are ubiquitous in garbage, is the leading hypothesis.

One dioxin compound—2,3,7,8-TCDD—is sometimes ominously called "the most toxic synthetic chemical known." In animal tests, it has been found to be 500 times more potent than strychnine and at least 10,000 times more potent than cyanide. The toxicity of dioxin, however, varies from species to species; whether dioxin is quite so toxic to humans remains unanswered. Still, immune system damage from dioxin exposure has been observed in a broad range of animal experiments, and there is little disagreement that TCDD, at least, is a potent cancer-causing agent.[42]

The greatest health risks come from older incinerators. Most apartment house garbage burners, like those in New York City, are rudimentary devices, with antiquated designs that burn inefficiently. They lack modern antipollution equipment. They are operated by untrained personnel. They have low stack heights. And their pollutants are usually emitted in densely populated neighborhoods. Many of these characteristics are also true of the city's scores of existing hospital incinerators and the three older municipal incinerators.

Air emissions from existing incinerators are most threatening to persons living or working in the immediate vicinity. City and state officials, none too anxious to uncover the extent of this nettlesome problem, have undertaken little on-site air monitoring of apartment house incinerators in New York City. Monitoring of hospital incinerators has been nil. But reports from elsewhere suggest that many on-site incinerators do not even meet the old and discredited federal standard governing particulate emissions from incinerators, adopted nearly two decades ago. And according to some experts, older, on-site incinerators are emitting more toxic pollutants per pound of refuse burned than modern, well-run hazardous waste burners. It is hardly a surprise that the Environmental Protection Agency projects maximum cancer risks from exposure to emissions from existing incinerators may be as high as 1 in 1,000 to 1 in 10,000—among the highest air quality risks identified by the federal agency. [43]

What about the more modern "resource recovery" incinerators? Sanitation officials maintain that these facilities don't pose the same risks as older, on-site garbage-burning plants. There is some truth to this assertion.

Modern pollution control equipment, if operated properly and well maintained, can significantly reduce air discharges of most pollutants from uncontrolled garbage-burning incinerators. One such device is called a "scrubber"—really a variety of different systems, all of which use some kind of alkaline spray to neutralize and remove acidic gases. The scrubber's companion on the proposed Brooklyn Navy Yard incinerator is the fabric filter, or "baghouse"—it resembles a series of very large vacuum cleaner bags, which can capture particulates of varying size (and most metals and organics that adhere to them). The Environmental Protection Agency reports that the best pollution controls (which include alkaline scrubbers and fabric filters) when combined with proper incinerator operation, should remove 90 percent of hydrogen chloride (70 to 90 percent of sulfur oxides), 99 percent or more of heavy metals (except mercury), and more than 95 percent of organics including dioxin. [44]

But such assurances have not quelled public concerns. And lingering suspicions as to long-term consequences of large-scale garbage burning in New York City seem justified. For one thing, garbage-burning technology is complex and understanding of issues like dioxin formation and control is still incomplete. In addition, it doesn't take much going wrong to result in sizable variations

TABLE 1.2 PROJECTED AIR POLLUTION EMISSIONS FROM THE PROPOSED BROOKLYN NAVY YARD INCINERATOR

Pollutants	Tons per Year
Particulates	161
Sulfur dioxide	1,189
Nitrogen dioxide	2,972
Carbon monoxide	368
Non-methane hydrocarbons	66
Lead	15
Sulfuric acid	92
Hydrogen chloride	537
Zinc	28
Mercury	5
Formaldehyde	27
Others	*

New York City Department of Sanitation

* Lesser amounts of other pollutants will also be emitted, including cadmium, chromium, copper, nickel, arsenic, selenium, beryllium, fluoride, polyaromatic hydrocarbons, polychlorinated biphenyls, polychlorinated dibenzo-p-dioxins, polychlorinated dibenzo furans, tetrachlorinated dibenzo-p-dioxin, and 2,3,7,8-TCDD.

in an incinerator's burning efficiency. Even if only a small portion of the total waste load experiences less than optimum combustion, air emissions can jump.[45]

Moreover, the city's track record in operating major public works facilities (such as old incinerators and sewage treatment plants) can best be described as uneven. And it is actual day-to-day operating practices, as much as the pollution control devices installed, that will determine air contaminant discharges from the new incinerators.[46] Don't count on blitzkrieg enforcement if garbage incinerator emissions are worse than expected, either. City and state officials have shown a distinct reluctance to close down incinerator operations (there are obvious practical difficulties in such a step), regardless of evidence as to air pollution discharges.

Finally, even if the new incinerators perform as promised, they will be major league emitters of many pollutants. The Brooklyn Navy Yard incinerator, for example, is expected to discharge annually more than 160 tons of particulate matter, 65 tons of hydrocarbons, 2,972 tons of nitrogen dioxide, 1,188 tons of sulfur dioxide, 368 tons of carbon monoxide, 27 tons of zinc, 14 tons of lead, 5 tons of mercury, and various amounts of many other pollutants, including dioxins and furans, according to the project's final environmental impact statement. These pollution levels would make even one garbage-burning incinerator the largest new source of air pollution in years to set up shop in New York City.

ASH RESIDUE

As if air pollution weren't enough, there's also the vexing problem of incinerator ash. Not everything dumped into a solid waste incinerator burns. Heavy metals like lead and cadmium don't. Nor do objects made of iron and glass. Whatever doesn't burn ends up either smoking out of the stack or being collected in the incinerator as ash residue.

The ash problem is no small matter. At the proposed Brooklyn Navy Yard facility, city sanitation officials hope to burn 3,000 tons a day of trash in the incinerator's large furnaces. If they do, they will be left with 800 tons of bottom ash and 135 tons of fly ash every day.[47] ("Bottom ash" is the charred, unburned material that remains on the incinerator grates after leaving the furnace chamber. "Fly ash" includes the solid substances captured in the incinerator's pollution control device.)

It created quite a stir in early 1987 when the Environmental Defense Fund (EDF) suggested that incinerator ash had the properties of toxic waste. Some government officials and incinerator advocates were quick to criticize the claim, challenging the appropriateness of existing test procedures. They opposed classifying incinerator ash as "hazardous," in part because that would mean higher disposal, liability, and insurance costs. That in turn would complicate, if not cripple, plans for rapid growth in the number of operating incinerators. But official test results have largely corroborated the EDF allegations. A 1987 Department of Environmental Conservation test of ash from six incinerators in New York State found that more than half of all samples tested had exceeded federal lead and cadmium levels for what constitutes hazardous waste. Ash may also contain dangerous levels of dioxins.[48] Ironically, as incinerator air pollution devices have become more effective in capturing metals and other contaminants, the toxicity of fly ash residue has increased.

Incinerator ash is not now being disposed of as a hazardous waste in New York City. Nor is it even being treated as a "special waste"— a new category created by the state, which imposes slightly more stringent requirements than those applicable to raw garbage. Ash from the three existing municipal incinerators is being barged to the Fresh Kills landfill and buried with ordinary waste. City plans to dispose of ash from the proposed Brooklyn Navy Yard plant at Fresh Kills has held up issuance by the state of the Navy Yard construction permit for months.

Little hard data are available on the health and environmental impacts of exposure to incinerator ash. But more than enough is known to raise a warning flag. The metals and dioxins, as previously discussed, pose known hazards to human health and environmental quality. Following incineration, the metals often adhere to microscopic particles, which can easily be dispersed by wind or water. Public exposure to the toxic ash residue can occur during on-site handling, storage, transport, landfill handling, and burial. At every step, both air-borne and water-borne dispersal is possible.

The Law

When garbage-burning proponents announce that a proposed or operating incinerator is meeting all existing regulatory standards, they are seeking to soothe public fears as to the safety of burning trash. Those who are familiar with the patchwork of environmental rules in this area, however, are hardly comforted by such assurances. Neither federal, state, nor city law has yet to set down a comprehensive regulatory framework to protect public health and the environment from the air pollution and ash residue risks of incineration.

There has been little more than a flicker of national regulation in this area. This reflects the historic view that solid waste disposal is primarily a local responsibility. The U.S. Environmental Protection Agency's air emissions standards for incinerators, one of the few federal incursions, are nearly two decades old and hopelessly out-dated.

If the 1980s has seen a shift in environmental regulatory power from Washington, D.C., to the states, solid waste is a case in point. In the absence of strong federal action, New York and other states have sought to fill the gap. The State Department of Environmental Conservation has succeeded in part. In 1988, the department adopted new rules establishing, among other things, numerical emission standards for air pollutants (including dioxin) from new incinerators. The rules do not cover existing incinerators, a major gap, but place important limitations on pollutants from new garbage-burning plants. The rules also establish tighter landfilling requirements for the disposal of incinerator ash. But they do not impose requirements equivalent to those for landfilling of haz-ardous waste.[49]

The city's air pollution and sanitation codes give the munici-pal government broad authority to set standards and abate air pollu-tion or ash contamination problems. But, for the most part, they do not specifically mandate that tough environmental rules be adopted.

Government Action

There has recently been a stirring of activity on the incinerator issue in Washington, D.C. At EPA headquarters, new incinerator rules are finally in the works. And on Capitol Hill, pending legisla-tive proposals would establish limits on garbage-burning air emis-sions and provide for ash handling.[50] But don't count on the final legislation to impose tougher requirements than the moderate in-cinerator rules already in place in New York State.

In Albany, meanwhile, the Department of Environmental Con-servation hopes soon to issue regulations governing emissions from existing incinerators. If the agency stays true to form, these new rules will move in the right direction, but fall short of providing the full measure of needed relief.

Good news sometimes comes from unexpected quarters. Just when it seemed as if government actions on the incinerator front would always be presented in watered-down fashion, an exciting piece of legislation surfaced from the New York City Council. Under the leadership of former Environmental Protection Committee chairman Sheldon S. Leffler, the council enacted legislation in 1989 directing owners of the 2,200 remaining apartment house incinerators to cease burning by mid-1993. The time period provided by the council is longer than necessary. But the long-awaited shutdown of this class of incinerators is one of the more welcome environmental developments in New York City in years.

Which brings us to 1990 and New York City's newly elected mayor, David Dinkins. The former Manhattan borough president ran on a platform that, among other things, called for full-throttle recycling and a moratorium on the construction of any new garbage-burning incinerators. He wants to see the results of the Sanitation Department's on-going waste composition study and its new solid waste management recommendations first. Here's one official, at least, who seems to be asking some of the right questions about garbage disposal in New York City. But as political scientists will tell you, campaigning is one thing and governing is another. The weighty responsibility for designing a workable solid waste battle plan could be the first big environmental test for the new Dinkins administration.

PART 2

Waterways and the Coast

Many New Yorkers have written off the city's waterways. Most of our rivers and bays are not fit for fishing or swimming. For both sewage and toxic contaminants, pollution levels frequently exceed environmental and public health standards. And too often our waterways are serving simply as marine highways. Yet, public clamor over these remarkable facts has generally been muted, at least until recently. Even with increased political attention following highly publicized beach closings in the summer of 1988, few city residents expect much to change.

To be sure, news from the waterfront is not all bad. Sparked by congressional mandates and court orders, the city is nearing completion of a decades-long struggle to construct or rehabilitate more than a dozen sewage treatment plants that now ring the five boroughs. And city and state officials have begun what may prove to be an equally ambitious campaign aimed at curbing the flow of industrial and commercial toxins into New York's waterways.

Does this mean that New Yorkers will soon be able to swim, boat, and fish in unpolluted waters? In a word, no. Existing laws and regulations leave major gaps in the control of sewage and toxic

47

contaminants. And funding for new pollution abatement and enforcement activities is unlikely to match future needs.

At the same time, new battle lines are being drawn over another long-ignored water resource—the city's 578-mile coastline. Up and down Manhattan's East and West Sides and in less publicized spots throughout the five boroughs, developers have begun to sketch out a new wave of residential and commercial projects that could permanently alter the city's shoreline. But citizen and neighborhood groups are lining up against many of these proposals in what are already heated conflicts over coastal land use and the appropriate size and scale of waterfront development.

SEWAGE POLLUTION

Background

Sewage historically has been the city's single largest source of water pollution. Through the turn of the twentieth century, virtually all of the city's raw sewage was dumped directly into New York's rivers and bays. By the 1920s, harbor surveys revealed water quality in decline as a result of this practice. And as New York's population and industrial base grew, city planners recognized that remedial action would be necessary.

The city's first systematic effort to reduce raw sewage flows came in 1935, with the construction of a sewage treatment plant on Brooklyn's Coney Island shore. Over the next three decades, additional facilities were built in all five boroughs. And by the end of the 1960s, New York City sewage plants were providing some degree of treatment for up to 80 percent of the sewage wastes discharged during dry weather periods.[1] Monies from the federal Water Pollution Control Act of 1972 spurred the upgrading of existing sewage treatment works and the construction of two additional plants.

Today, New York's 14 sewage treatment plants can, during dry weather, handle virtually all of the city's daily 1.7 or more billion gallons of raw waste. Not all of these facilities meet government requirements for treatment plant efficiency. Nevertheless, the two leading indicators of sewage pollution—bacteria counts and dissolved oxygen levels—have declined significantly since the early 1970s. For example, levels of dissolved oxygen in 11 sections of New York Harbor improved, increasing by 30 percent as measured by the city Department of Environmental Protection from 1972 to 1984.[2]

Despite all the hoopla over treatment plant construction, sewage pollution remains a serious problem for New York's waterways. The main offender is often referred to as combined sewer overflow, or CSO. It occurs because a single network of piping in most parts of the city carries both sewage wastes and stormwater runoff. When it rains, the volume of combined sewage and rainwater can exceed the intake capacity at sewage treatment plants. As a result, large volumes of combined rainwater and sewage (along with assorted toxins) often bypass treatment plants and flow directly into New York's rivers and bays. As construction and upgrading of the city's sewage treatment plants progress, CSO has emerged as the single most important unresolved water pollution problem in New York.

Sewage Impacts

Wading into New York's waters is easier said than done. New York City Health Department officials have permanently banned swimming in the Hudson River, Jamaica Bay, and along vast stretches of the waterfront, in part because of the presence of high levels of sewage-related bacteria. Similar concerns have prompted the department to designate other popular areas, including Graham Beach on Staten Island, as "not recommended" for swimming. Sewage pollution has not spared the city's major beachfronts. The city Health Department has issued a standing warning against swimming for at least two days following heavy rains at the northernmost Staten Island beaches (including South and Midland), the southernmost Bronx beaches (including Locust Point and Little Neck Bay), and the westernmost Coney Island beaches (including Seagate and Coney Island). Combined sewer discharges following recent summer rainfalls raised water bacteria counts to unsanitary levels, in some cases as much as 500 percent or more above normal dry weather concentrations.[3]

Sewage has also had a dramatic impact on shellfishing. At the turn of this century, the city supported a prosperous shellfish industry. By the 1920s, however, sewage contaminants had been largely responsible for the closure of oyster and other shellfish beds in virtually all of the city's rivers and bays. Today, coliform bacteria concentrations in New York City waters remain so high that the state continues to prohibit, in most city waters, the taking of clams, crabs, and other shellfish. Despite these restrictions, some New Yorkers apparently still harvest from contaminated city waters.[4]

Shellfish have not been the only victims. The presence of raw or inadequately treated sewage in waterways consumes oxygen necessary to sustain many species of marine life. Severe oxygen drops in regional waters, caused in part from sewage-related wastes, have resulted in occasional fish kills. Less serious declines in oxygen levels contribute to more subtle adverse impacts on fish and other organisms that dwell in New York City's often murky waters.[5]

The Law: Sewage

As farfetched as it now sounds, Congress sought in 1972 to curb the flow of sewage and other pollutants into the nation's waters by the mid-1980s. One of the primary goals of the 1972 Federal Water Pollution Control Act (later renamed the Clean Water Act) was the restoration of waters to fishable and swimmable quality. For sewage pollution, the act established a two-pronged approach. First, it required all localities to upgrade existing municipal sewage plants to secondary treatment by 1977. (Secondary treatment is defined primarily as the removal of 85 percent of sewage pollutants.) Congress later extended this deadline for certain municipalities, including New York City, until 1988. Second, it directed the states to classify their water bodies based upon existing or anticipated uses and to reduce pollution discharges so that the waters would be safe for each of these uses.[6]

Sewage Treatment Plants: Government Action

New York produces more raw sewage than any other city in the nation. Not surprisingly, it also has the largest number of municipal treatment plants. After years of delay, 14 such facilities are now operating throughout the five boroughs. Among them is the new $1.2 billion North River Water Pollution Control Plant on Manhattan's Upper West Side. It is designed to treat 170 million gallons a day of raw sewage, slightly more than the amount that had previously been dumped directly into the Hudson River.

While the city has made substantial progress in constructing sewage plants, levels of treatment vary considerably. During the first half of 1989, only four city plants were regularly meeting the Clean Water Act's 85 percent removal requirement for sewage pollutants. Three of the most troubled plants (Coney Island, Owls Head, and Newtown Creek) consistently failed to remove either 85 percent of suspended solids, 85 percent of oxygen-demanding matter, or both. Three other plants (Bowery Bay, Wards Island, and Jamaica) did not meet the 85 percent removal requirement in four or more months during that same period. Regarding the two newest plants, North River is removing only 20 to 70 percent of its sewage wastes, but under court order is expected to achieve the more advanced secondary treatment level of 85 percent by 1991. The Red Hook facility was projected to have reached secondary treatment as 1989 came to an end.[7]

What accounts for these performance problems? Like the city's bridges and its transit system, New York's sewage treatment plants have been beset by problems of age and deferred maintenance. The city reports, for example, that the Owls Head and Coney Island treatment plants have "severe structural and operational problems due to their age." And the Interstate Sanitation Commission, a tri-state environmental agency, in documenting treatment plant maintenance woes in 1986, found sewage pumps and other major equipment to be inoperative at a number of city facilities. On occasion, the equipment shutdowns had lasted for several years.[8]

The 1987 Clean Water Act amendments, passed over then-president Ronald Reagan's veto, offer a ray of hope. As much as $650 million in new federal grants will be available to New York City for sewage-related work over the next several years. More than $900 million in federal loans may also be forthcoming. The primary beneficiaries of this capital infusion include some of the city's oldest and least efficient sewage plants. Work is already underway at the ailing Coney Island and Owls Head facilities.

FIGURE 2.1: NEW YORK CITY'S 14 SEWAGE TREATMENT PLANTS

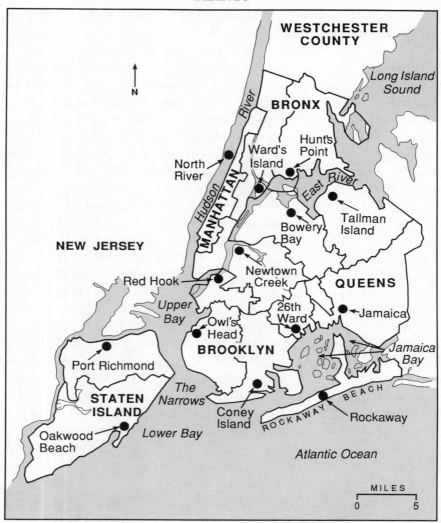

New York City Department of Environmental Protection

But the outlook is not all sunny. Reconstruction (and thus consistent attainment of secondary treatment levels) at the Coney Island, Owls Head, and Newtown Creek facilities is not expected to be completed until the mid-1990s. It took a lawsuit by New York State's attorney general Robert Abrams and the state Department of Environmental Conservation to lock in even these dates. Maintenance cutbacks at sewage plants have also jeopardized the city's ability to comply with secondary treatment requirements. City Hall's fiscal 1987 budget, for example, shaved 90 employees and over $2.5 million from the Department of Environmental Protection's plant maintenance program. By the city's own forecast, these reductions "will result in an increased backlog of corrective maintenance projects and a reduced level of preventive maintenance work at the plants." Budget changes in the years since then have hardly turned this situation around.[9]

Are 14 treatment plants enough to handle New York City's daily sewage load? Some evidence suggests that the answer is no. During the first 6 months of 1989, 5 of the city's 14 plants were operating at least half of the time at or over their design capacity. The most glaring example is the Wards Island treatment facility, which serves households and businesses on Manhattan's Upper East Side and portions of the Bronx. During some months of 1989, it was receiving daily sewage flows that averaged more than 145 percent over its existing design capacity.[10] These overloads will likely take their toll on the durability and efficiency of the city's expensive treatment plant equipment. And the issue is sure to heat up as plans for new commercial and residential development threaten to add additional millions of gallons of sewage to already overburdened treatment works. (In fact, the city is now talking about expansions of the Newtown Creek and Wards Island plants.)

Combined Sewer Overflow

Decisions made a century ago by city sanitation planners continue to haunt ongoing sewage cleanup efforts. In the late 1800s, engineers began constructing a citywide network of pipes to carry most of New York's household wastewater and stormwater runoff in a combined system. These planners could not foresee that mixing wastewater and stormwater in a single pipe would substantially complicate pollution abatement in decades to come.

Today, when it rains or snows, stormwater rushes through the city's combined sewers at volumes 2 to 20 or more times the dry weather flow of household wastewater alone. These larger volumes, if not diverted, would overwhelm treatment plants that were primarily designed to handle sewage wastes. A system of bypass valves diverts the excess stormwater-sewage mixture away from treatment plants to avoid flooding those facilities. But as a result, untreated sewage (along with stormwater runoff) pours directly into the city's waterways from nearly 500 open drainpipes and outfalls. This phenomenon, often referred to as combined sewer overflow (CSO), is the city's number-one sewage problem today.[11]

Heavy rains and snow are not the only culprits. The bypass valves that divert wastewater and stormwater during wet weather sometimes malfunction. As a result, raw sewage is often, although inadvertently, channeled into city waterways even during dry weather. Additionally, periodic mechanical problems at intermediate pumping stations and other sewer works also send raw wastewaters gushing from time to time.

Looking for big numbers? During an average seven-hour rainstorm, as many as 560 million gallons of combined sewage and stormwater are diverted directly into the city's waterways. And even after light rains that do not overwhelm treatment plant capacity, valve malfunctions in combined sewers funnel an additional 35 million gallons of wastewaters (over seven hours) into New York's rivers and bays. Among the areas most affected are city beachfronts, as well as tributaries and small inlets such as Flushing Bay in Queens and Brooklyn's Gowanus Canal and Newtown Creek. The city's Department of Environmental Protection has concluded that combined sewer overflow is "the major source of contamination in New York City's waters."[12]

It takes more than technical know-how to solve most environmental problems. Legislative mandates and statutory deadlines are usually necessary. CSO is a case in point. More than 20 years ago, environmental engineers drafted a citywide strategy to confront the CSO issue head-on. At the heart of this plan was the construction of large holding tanks to capture and store sewer overflows and send these wastes to nearby treatment plants after stormwater volumes had subsided. In the late 1960s, one such facility was opened at Brooklyn's Spring Creek, on the north shore of Jamaica Bay. Despite that plant's apparent success, only one other similar facility— another pilot project at Fresh Creek on Jamaica Bay—has ever

O'Brien and Gere Engineers, Inc.

ONE OF 500. *With the construction or upgrading of 14 municipal treatment plants nearing completion, the discharge of raw sewage from combined sewer and stormwater pipes has become perhaps the city's number-one water quality problem. Nearly 500 outfalls, like the one pictured above, pour sewage into city waterways, primarily after heavy rainfalls.*

materialized. Funding priorities and statutory deadlines that favored the upgrading of sewage treatment plants over CSO abatement help explain why.[13]

Until now, CSO pollution control in New York City has slid through the cracks. City efforts have primarily been confined to monitoring and data gathering. Only recently has the Department of Environmental Protection begun to zero in on the problem in three of the most affected water bodies. Dredging of sewage sediments and repair or replacement of faulty sewage valves have commenced at Flushing Bay. And plans for the installation of a 40-million-gallon sewage holding tank for the Flushing area are now

FLUSHING BAY: LIVING UP TO ITS NAME

Have newly constructed treatment plants solved the city's sewage problem? Just ask the residents of Flushing, Queens. In their neighborhood lies Flushing Bay, one of the borough's most valuable natural resources. It is also one of the most polluted.

Flushing Bay is located in northern Queens, flanked on one side by LaGuardia Airport and by the College Point neighborhood on the other. Jutting inland from the bay is the narrow Flushing Creek, sometimes called Flushing River; it connects the bay with Flushing Meadows–Corona Park, the borough's largest and most-visited public recreation area.

Raw sewage discharges are the bay's number-one environmental problem. Fourteen combined sewer outfalls drain into Flushing Bay and its adjacent creek, contributing 60 million gallons or more of sewage wastewaters during an average seven-hour rainstorm. The problem is not merely a wet weather phenomenon. Even during dry weather, faulty sewer regulators and other mechanical problems can channel several hundred thousand gallons a day of wastewaters, intended for nearby treatment plants, directly into the bay and creek.[14]

Odor problems, aesthetic problems, water quality degradation—they can all be found in Flushing Bay. In fact, official monitoring reports reveal that the bay is consistently among the city's most severely affected water bodies when it comes to sewage-related pollutants. Bacteria counts and oxygen levels—two common measures of sewage pollution—frequently violate state water quality standards designed to protect public health and aquatic life. And the buildup of odor-producing sewage sediments near outfall pipes is a recurring problem for bay visitors and nearby residents.[15]

Government plans to clean up the bay's sewage pollution are still largely on the drawing boards. The most ambitious project proposed by the city's Department of Environmental Protection calls for the building of a 40-million-gallon underground holding tank in Flushing Meadows Park. The tank would capture and store a portion of the wet weather sewage overflows until these wastewaters could be handled by nearby treatment plants. The department has also begun to dredge sections of the bay and creek to remove sewage sediments. For the residents of Queens, these projects can't come fast enough.

FIGURE 2.2: FLUSHING BAY

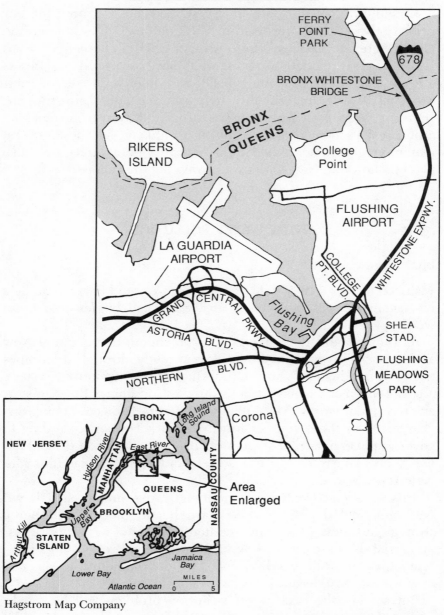

Hagstrom Map Company

pending. So, too, are proposals for CSO control at Brooklyn's Gowanus Canal and Paerdegat Basin in Jamaica Bay.

Lack of progress in controlling CSO here has mirrored the low priority the issue has received in Washington, D.C. Although the Clean Water Act requires the control of CSO discharges, EPA has issued no clear regulations saying what those controls should be. As a result, state environmental officials have been left to devise their own control programs with little help from the federal government. In recent years, New York State has directed the city to get cracking on CSO abatement strategies. But it is unlikely that these general planning requirements and limited government funds will curb CSO pollution in New York's waterways anytime soon.

TOXIC WATER POLLUTION

Background

Fish don't vote. If they did, New York's rivers and bays would be a lot cleaner. Not only would there be less sewage, but fewer toxins as well.

As things stand now, however, toxic chemicals pour into New York's harbor every day. Some flow directly through drainpipes from major industrial plants. Others make their way into city waters after an indirect journey from businesses and homes through sewage treatment works. More than 60 different toxins are being discharged from these sources into New York Harbor, according to government estimates that are probably on the low side.[16] Contaminants also run freely off city streets and municipal landfills into New York's waters.

This situation has not gone unnoticed. Government officials and environmental decision-makers have begun piecing together a strategy for ending the toxic onslaught. Many of the city's largest industrial dischargers, for example, have been brought under the regulatory umbrella of federal and state environmental laws passed over the last two decades.

But solving the toxic water pollution problem will be tough. Many toxins are extremely persistent and can pose long-term environmental and health risks for some time after their initial discharge. Moreover, toxic pollutants make their way into New York Harbor not only from city sources but also from discharge pipes and runoff upstate and in New Jersey. And even fallout of airborne contaminants would have to be listed if you were preparing a toxic water pollution roster. Securing environmental change is difficult when the political constituency in favor of change is perceived as weak. This helps further explain why the last two decades have seen greater progress in controlling the more visible sewage problem than the insidious toxic threat to New York City's rivers and bays.

Toxic Water Pollution Impacts

Almost 20 years after passage of the Federal Water Pollution Control Act, toxic pollutants in New York City waterways are still contaminating local fish and shellfish populations. Polychlorinated biphenyls (PCBs) are among the chief villains. These potential human carcinogens were dumped for decades into the Hudson River, primarily from two General Electric industrial facilities near Albany. They are still being detected in fish, often at levels well in excess of federal guidelines.[17]

In response, state officials have taken extraordinary action. They have issued warnings against public consumption of striped bass, carp, bluefish, and other species taken from local waters. And they have banned the commercial taking of nine fish species from the Hudson River and throughout New York's coastal region. No wonder. Scientists now believe that ingestion of fish may be the primary nonoccupational source of exposure to PCBs and other toxic contaminants. Yet, many New Yorkers remain unaware of these dangers, and some recreational anglers are still consuming fish hooked from New York City's waters.[18]

You might be surprised by some of the other substances turning up in fish these days. Cadmium, mercury, and other heavy metals are among the unwelcome toxins that have been found in fish and shellfish taken from nearby water bodies. So, too, have organic compounds such as dioxins.[19] Still, environmental experts know relatively little about toxic water contamination in New York. No comprehensive analysis of the presence of toxins in the city's marine life has ever been completed. Nor has there been any full-scale assessment of the impacts of toxic water pollution on New York's environment or public health.

What is known about water-borne contaminants, however, is hardly reassuring. Concentrations of four toxic metals (lead, zinc, copper, and nickel) in New York City waterways consistently exceed state water quality standards designed to protect aquatic life. In total, six of nine metals monitored by the city's Department of Environmental Protection are violating these standards. And, significantly, with the exception of lead, levels of heavy metals in local waters have not changed appreciably, if at all, since regular city monitoring for these toxins began in the early 1970s. Data on organic compounds are more skimpy; suffice it to say that such chemicals are regularly detected at surface water monitoring stations, sometimes in excess of state water quality limits.[20]

Still, evidence suggests that loadings of some toxic pollutants have declined. For example, discharges of several of the most notorious chemicals, including DDT, PCBs, and lead, have dropped from peak levels of earlier decades, due largely to government restrictions and changes in industrial processes. Because such toxins linger in harbor sediments, however, the impacts of past dumping will continue to be felt in New York City's waters for years to come.[21]

How do New York's waterways stack up against other areas in terms of toxic water contamination? Such comparisons are not easy. But at least one survey by the National Oceanic and Atmospheric Administration has concluded that levels of toxic contaminants in nearby Raritan Bay (which separates Staten Island from New Jersey) make it one of the most polluted harbors in the country.[22]

Who's Polluting New York Harbor?

DIRECT DISCHARGERS

When one thinks of water pollution, what comes to mind are open drainpipes from which wastewaters cascade into rivers and bays. The image may be dramatic, but New York has its share of these so-called direct dischargers. The New York State Department of

Environmental Conservation classifies 27 facilities here as "major" direct dischargers. Fourteen of these are the city's sewage treatment works, ten are utility plants, and three are industrial firms. There are also roughly 90 other city sources that directly discharge lesser amounts of pollution into surrounding waters. Petroleum terminals and storage facilities are perhaps the single largest category in this grouping, with the exception of smaller sewage dischargers not presently hooked up to municipal treatment plants.[23]

Are these direct dischargers a major threat when it comes to toxic pollution? Limited data make assessments risky. Only a dozen or so New York City–based sources hold state permits to dump toxic contaminants. But that does not include the city's sewage treatment plants, probably the largest local source of water-borne toxins. Nor does it encompass facilities releasing pollutants such as oils and grease. Nonetheless, while some direct dischargers may be responsible for localized problems, it is likely that these facilities collectively are not the primary source of toxic contamination in New York waterways.

INDIRECT DISCHARGERS

A roundup of the city's toxic water polluters would have to include more than the usual suspects. The largest category of toxic dischargers in New York do not dump directly into the city's waters. Instead, they pour their wastes into the sewer system. No precise head count of these seemingly clandestine dischargers exists. But city officials have so far identified just over 800 industrial and commercial firms that may be indirectly dumping toxic pollutants in this fashion. Of these, approximately 420 are electroplaters or metal finishers. Paint and ink formulators, metal molding and casting shops, pesticide makers and pharmaceutical manufacturers help fill out the list. The official figures probably underestimate the total number of indirect industrial and commercial toxic polluters.[24]

The situation would not be so bad if the city's sewage network could adequately treat toxins streaming through the system. It can't. In fact, every day more than 7,000 pounds of metals enter city sewage treatment plants. (This figure does not include metals that are carried directly into our waterways from street runoff.) The

Eric A. Goldstein

NEW YORK'S LARGEST TOXIC WATER POLLUTER? *Surprisingly, most toxic water polluters in New York City do not dump directly into rivers and bays. Instead, these industrial and commercial firms discreetly dispose of their wastes in the sewer system. But New York City's sewage treatment plants, such as the Newtown Creek plant pictured above, are unable to neutralize most toxic contaminants. And until individual firms adequately pretreat their wastes and household and small business discharges are controlled, city sewage works will continue to receive more than 7,000 pounds of toxic metals every day.*

metals either pass through the plant into surrounding waters or end up in sewage sludge, which is presently dumped 106 miles off-shore. Zinc leads the way, according to city data, followed by copper, lead, chromium, nickel, cadmium, and mercury. These discharges include not only industrial wastes, but also those from households and small businesses, which have been singled out as significant players in some aspects of the toxic metal dumping game.[25]

URBAN RUNOFF

Want to observe toxic water pollution firsthand? Just wait until the next time it rains. Then head to the nearest street corner, where motor oils and fuel, identifiable by their iridescent sheen, as well as particulate matter and street dirt, are being flushed into storm sewers. Or visit other observation posts, including airports, construction sites, and even residential lawns. They all are sources of toxic water pollution, when rain washes contaminants (oils and grease, pesticides, fertilizers) into rivers and bays in the phenomenon known as urban runoff.

Scientists only recently have begun to piece together an accounting of the pollution loads from urban runoff. Available information, admittedly sketchy, suggests that as much as 3,000 pounds a day of zinc, lead, copper, chromium, and other heavy metals are flushed directly into surface waters from city streets, stormwater and combined sewer pipes. Smaller amounts of organic chemicals, such as benzene, toluene, and PCBs, are also washed into surrounding water bodies. [26]

What distinguishes urban runoff from other sources of toxic water pollution is that it does not always flow from a single pipe or drain, or even from a particular geographic area of the city. A clearer example of uncontrolled pollution would be hard to find.

The Law: Toxic Water Pollution

The Clean Water Act spells out three separate programs for controlling toxic contaminants that pour into the nation's waterways. The first applies to direct discharges. It requires these facilities to secure permits establishing allowable toxic flows. In theory, each permit is based upon the use of best available pollution control technology and the quality of surrounding waters. [27]

The second set of national rules applies to indirect dischargers— those disposing of toxic wastewaters via the sewer system. This category of polluters is subject to the act's so-called pretreatment requirements. These rules are designed to safeguard the operation of sewage treatment plants, protect their workers, insure the quality of surface waters, and prevent contamination of sewage sludge. Under these provisions, major industries, such as electroplaters, must install technology-based pollution controls at their facilities to treat wastes on-site, before disposing of them in city sewers. Industrial and commercial dischargers in some municipalities must also meet locally adopted limits on toxic pollution.[28]

The third pool of toxic pollution is urban runoff. The Clean Water Act requires that every state identify waters poisoned by runoff and devise strategies to control the problem. But the 1987 amendments do not mandate immediate action on this admittedly thorny problem. Instead, they direct states to study localized runoff trends and to prepare four-year management plans. The new law also requires that New York City and other urban areas secure stormwater discharge permits by 1991. These permits must, among other things, set in motion local programs to reduce stormwater pollution to the maximum practicable extent.[29]

Government Action: Toxic Water Pollution

Environmental statutes are big on lofty pronouncements. The Clean Water Act boldly proclaims that "the discharge of toxic pollutants in toxic amounts be prohibited."[30] But legislative intent is one thing; actual implementation is another.

Toxic water pollution in New York City exemplifies the problem. Despite the existence of sweeping legislative goals, cleanup efforts are nearing a stalemate. Government restrictions and regulatory programs have apparently blocked major new declines in regional water quality. But it is unlikely that existing initiatives alone will cleanse New York's waterways of toxic pollutants as envisioned by Congress.

Take indirect dischargers. At first glance, it appears as if the city is on the road to satisfying the broad Clean Water Act goals. After years of delay by national and local regulators, a federally approved New York City pretreatment program is now in place governing major industrial and commercial firms that dump toxic wastes into city sewers.[31]

Peeling back the outer layer of the city's pretreatment program, however, reveals a somewhat different picture. Of the more than 800 firms city officials have identified, they are only regulating about 275; they have plans to get to the others, but have assigned them a lower priority. In addition, they have yet to investigate as many as 2,000 other business establishments, some of which may also be improperly discharging wastes into the sewer network.[32]

JAMAICA BAY: AN EMBATTLED ESTUARY STRUGGLES TO SURVIVE

Jamaica Bay is perhaps the most spectacular stretch of New York City's 578-mile coastline. It is situated between Brooklyn's southeastern shore and the thin Rockaway peninsula. The bay's shallow waters and low-lying island marshlands sprawl over nearly 13,000 acres, more than 15 times the size of Central Park.

Like other New York waterways, Jamaica Bay is an estuary—a coastal area where inland freshwater and ocean saltwater meet—providing an excellent habitat for fish, shellfish, and smaller marine organisms. The bay also offers sanctuary to amphibians, reptiles, mammals, and over 300 species of birds. In 1972, large portions of Jamaica Bay's waters and shorelines were brought under the protective wing of the federal government, with their inclusion in the newly created Gateway National Recreation Area.

Jamaica Bay has struggled for decades against a long list of environmental pressures. Sewage pollution has been the most persistent enemy. The construction of four city-owned treatment plants has ended much dry weather discharge of raw sewage. But combined sewer overflow problems can funnel more than 100 million gallons of stormwaters and sewage into the bay area during an average seven-hour rainstorm.[36]

Toxic water pollutants also find their way into this marine environment. Metals (such as copper, nickel, and zinc) and organic chemicals pass unfiltered into the bay every day, largely from industrial and residential sources dumping these wastes into city sewers. Leachate from several New York City landfill sites perched along Jamaica's shore add PCBs, mercury, and other contaminants. And urban runoff from sources such as John F. Kennedy Airport on the bay's northeast corner wash additional pollutants, including motor oil and grease, into the bay.[37]

Sewage and toxic discharges have taken their toll on Jamaica Bay. High bacteria levels from raw sewage have contaminated shellfish beds. As a result, state officials have placed shellfish harvesting in the bay off limits. Heavy metals and organic pollutants such as PCBs have also been detected—in some cases at levels nearly 100 times state water quality standards. These contaminants frequently turn up in the bay's fish and other marine species.[38]

FIGURE 2.3: JAMAICA BAY

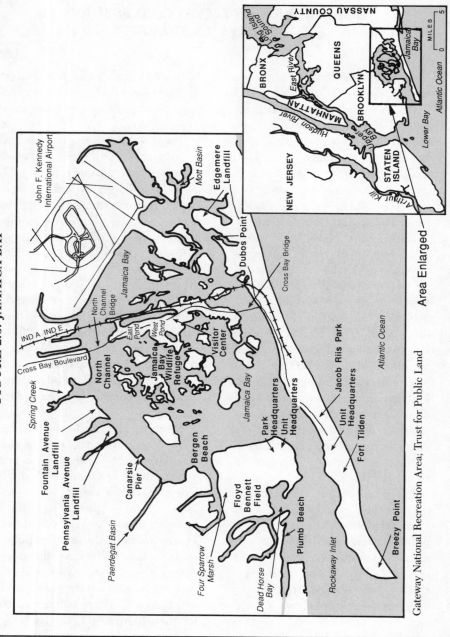

Gateway National Recreation Area; Trust for Public Land

At the turn of the century, Jamaica Bay was nearly double its current size and surrounded by an outer ring of saltwater wetlands and salt flats. Over the years, most of these fragile parcels have disappeared. Government officials allowed solid waste, sewage sludge, and dredge to be deposited in these former wetland areas. The result: parts of Jamaica Bay were turned into municipal refuse dumps and new land for development projects. Floyd Bennett Field (the city's first municipal airport and now part of Gateway) and nearby Kennedy Airport are among the legacies of these bittersweet land use decisions. To make matters worse, encroaching development, such as the Kennedy Airport runway, has disrupted the bay's tidal flushing with ocean water. The eastern bay, in particular, is now a reservoir in which sewage and toxic pollutants are often trapped.

These development pressures have not yet abated. Although city officials hold title to virtually all of the bay's remaining unprotected wetland and upland areas, they have targeted crucial locations, such as Paerdegat Basin, Spring Creek, the eastern shore of Fresh Creek, and various parcels in the Rockaways, for new residential and commercial development. To safeguard the bay and begin its restoration, environmental advocates have urged that the city instead set aside as natural habitat roughly ten miles of shoreline and 700 acres of wetlands and other ecologically sensitive adjacent properties.[39] The Parks Department and individual city officials have embraced this concept. And the Department's dedicated Natural Resources group has helped to acquire Dubos Point and Brant Point and to protect Floyd Bennett Field and several other parcels. But city development planners continue to resist movement in this direction.

Yet, what has surprised many observers is that despite these conditions, much of the bay's aquatic life continues to survive and adapt to its surrounding environment. More than 40 species of fish, ducks, shorebirds, and a rich diversity of smaller marine organisms inhabit this estuary. Even the endangered Ridley turtle has been seen paddling in the bay's troubled waters.

Jamaica Bay, then, is at a crossroads. The marine scientist and even the casual observer can find indications of unresolved environmental problems as well as signs of improving environmental health. Still unanswered is whether government officials will move to tip the balance in favor of preserving this unique urban sanctuary.

That's not the only trouble spot in the city's pretreatment program. City environmental officials have been hesitant to develop strict local limits on toxic pollutants that are dumped into city sewers. The federal Environmental Protection Agency has recently flagged the city's position as a "problem area," requiring "further action to bring the program into compliance with federal pretreatment regulations."[33]

How about direct dischargers? For firms dumping wastes directly into rivers and bays, the state, not the city, is New York's toxic gatekeeper. The state Department of Environmental Conservation (DEC) oversees the permitting of these toxic water dischargers. The program is known as the State Pollutant Discharge Elimination System, or SPDES.[34] As noted earlier, direct discharges from industrial and commercial firms are not the primary source of toxic water pollution in the harbor. And DEC has reportedly issued permits to all direct dischargers as required by law.

But state officials have not yet turned the string of individual permit decisions into a tightening noose of toxic control. DEC's water discharge permits are based solely on pollution control technology, and do not take into account pollution levels in the harbor, despite Clean Water Act language to the contrary.[35] The city's sewage treatment plants, surprisingly the largest source of toxic water pollution in New York, are a prime example. There is no presently available technology to control toxic pollutants at the plants themselves. But setting limits on toxic discharges from these facilities could prompt New York City to tighten requirements for industrial firms disposing of toxic wastes through the sewer system. Unfortunately, DEC's current permits place no lid whatsoever on the amount of toxins that pour out of municipal sewage works.

THE COAST

Background

It's open season on the city's waterfront. At least, that's the way some New Yorkers see it. They argue that unrestrained commercial and residential development is threatening irreplaceable shoreline parcels. Development serves no one, they add, if it means walling off the water's edge, overloading existing city services, and threatening the marine environment.

Many developers and some government officials see things differently, noting that much of the city's waterfront has fallen into decay and that the city needs the tax revenues and other economic benefits that would flow from new development. The real danger, they say, is not overdevelopment, but government red tape and citizen opposition that make siting new waterfront projects far too complicated and time-consuming.

Everyone agrees that the stakes are high. New York's 578-mile shoreline is twice as large as the waterfronts of Baltimore, Boston, Oakland, Philadelphia, San Diego, San Francisco, and Toronto combined. Of course, the fate of much of the coast is set, at least for now—two major airports, circumferential highways, 16 miles of beachfront, and countless commercial and residential developments from Brooklyn's River Café to Manhattan's Battery Park City are among the many existing uses that now dot the coastal landscape. But extensive areas of undeveloped or underdeveloped land, as well as docks and piers, are still up for grabs in all five boroughs.

Both the developers and the citizen warriors have a point. On balance, though, the critics of existing developing trends advance a more compelling case. To be sure, this does not mean all new waterfront development is inappropriate. But city officials must first put in place regulatory mechanisms for balancing legitimate waterfront development objectives with equally important environmental, recreational, and quality-of-life concerns. Until this occurs, expect fierce battles to continue along the city's coastal frontier.

Coastal Development: A Bird's-Eye View

Big changes are in the works along New York City's coast. In Manhattan alone, more than a dozen major waterfront development projects are now on the drawing boards. They include:

- *Trump City.* Located between 59th and 72nd streets on Manhattan's West Side, this proposed project envisions 11 residential buildings (including eight 60-story towers), two office buildings, and a 150-story skyscraper, along with 1.5 million square feet of total retail space.

- *Hudson River Center.* A 25-acre site, located between 35th and 40th streets, along West Street (opposite the Javits Convention Center), is slated to become a major commercial and residential center, which is likely to include a hotel, marina, heliport, and a new car tow-away pound; it would be built mostly on platforms over the Hudson River.

- *East River Landing.* A seven-million-square-foot, 23-acre development project along the East River just south of the Manhattan Bridge would include a new office, commercial, residential, and hotel complex, erected largely on platforms covering the East River.

- *Riverwalk.* This proposed 27-acre residential and commercial development project (1,888 residential units, a 245-room hotel, 240,000 square feet of retail and office space), situated between 16th and 24th streets on Manhattan's East Side, would entail the construction of 16 acres of platforms over what is now the East River.

Why worry about this kind of development? For one thing, some of these proposals would dramatically alter New York's rivers and shorelines as we know them today. The proposed Hudson River Center, East River Landing, and Riverwalk projects by themselves would add more than 60 acres of new land for residential and commercial development on decks over the Hudson and East rivers. And a potential northward expansion of Battery Park City could saddle the lower Hudson with dozens more acres of landfill, platforms, and floating structures. This would be only the latest wave of a centuries-old expansion of the city's coast. But the continued narrowing of New York's waterways could jeopardize the survival of fish and other marine species that make their home in the region's rivers and bays. Environmental reviews prepared during the battle over the now defunct Westway highway project, for example, revealed that proposed landfill in the Hudson River could have just this effect on its population of striped bass.

Waterfront development on a scale now being contemplated in Manhattan poses problems that extend beyond the waterways themselves. The Trump City proposal offers a sobering glimpse of what coastal parcels might look like in an era of unbridled development. Hardly any of the city's major environmental fronts would remain untouched.

Consider the following:

- Although it would not require landfilling or decking, Trump City would still tinker with the Hudson River ecosystem. The 33,000 new residents and employees (along with tens of thousands of shoppers and other visitors who would stream into the satellite city every day) would produce at least 2.3 million gallons of sewage daily, even as the nearby North River sewage treatment plant is already operating near or at its design capacity.

- Trump City's 7,300-car parking facility (the largest in Manhattan), its distance from existing transit lines, and its giant enclosed shopping mall would attract 25,000 vehicles to the area every day. Traffic along already congested West Side thoroughfares would increase sharply. And the new travel would spew additional air contaminants into an area that already violates national health standards for carbon monoxide and ozone.

- Trump City is expected to generate more than 23,000 new subway trips every day, straining existing transit services. Local stations, including the one at Broadway and 72nd Street, already suffer from serious rush-hour pedestrian bottlenecks.

- And Trump City's massive bulk (including eight 60-story residential towers and a 150-story skyscraper, the world's largest) could become both a physical and psychological wall that separates New Yorkers from the river and deprives them of light and air. The proposed office tower would at times cast its shadow across the Hudson into New Jersey, as well as over portions of Riverside and Central parks.[40]

Westpride Inc. and the Queens Museum

"I LIKE THINKING BIG. *I always have," says New York developer Donald Trump. "To me it's very simple. If you're going to be thinking anyway, you might as well think big." His proposed Trump City project, a model of which is pictured above, envisions the construction of a 150-story skyscraper—the world's tallest building—as well as a row of 60-story towers, one of the region's largest shopping malls, and parking facilities for more than 7,300 cars on Manhattan's West Side. The project typifies the runaway development that can occur in the absence of careful waterfront planning.*

The Trump City proposal and other development plans have rekindled a strong sense of citizen activism, especially in Manhattan. And they have sparked the formation of savvy and effective community groups, the best example being Westpride. These organizations, along with more traditional forces such as the Municipal Art Society (which brought the successful challenge to the 1985 Boston Properties proposal for the New York Coliseum site) and the Coalition for a Livable West Side, have already demonstrated their battle-readiness.

Coastal controversies are not limited to Manhattan. Of course, coastal development issues in the other boroughs do not always involve the same questions of density and open space. Still, residents of Brooklyn's Sheepshead Bay, for example, are greeting development plans in their neighborhood with mixed emotions. Some welcome a proposed $17 million plan, which includes a floating restaurant, retail shops, and residential condominiums on the bay. Others, including charter boat operators and some small businesses, fear that the proposed changes and lack of parking could erode the economically precarious fishing trade that has defined the character of the neighborhood for decades.

The issues are even more starkly drawn when coastal development knocks at the door of urban parkland. Gateway National Recreation Area is one of the city's coastal jewels. And plans to develop several sites in and around this area have not unexpectedly tripped the alarm among conservationists and concerned citizens.

Land use experts at the Trust for Public Land and the New York City Audubon Society have questioned the wisdom of city plans to develop critical sites buffering Jamaica Bay. Fourteen acres in Mill Basin, Brooklyn (adjacent to wetlands known locally as the Four Sparrow Marsh); 21 acres on Brooklyn's Paerdegat Basin (which feeds directly into Jamaica Bay); and a 10-acre peninsula in Queens (poking into the bay and located between Vernam and Barbadoes Basins in the Rockaways)—these are among the properties that have been targeted for residential, commercial, or industrial projects by the Public Development Corporation, New York City's lead agency for waterfront development. But conservationists argue that these and other city-owned sites should instead be designated as parkland to protect environmentally critical areas and the larger Jamaica Bay ecosystem. [41]

A recent controversy erupted over the fate of a portion of Jacob Riis Park, a popular 400-acre shorefront unit of the Gateway Recreation Area. A proposed National Park Service plan would have allowed private developers to build a 15,000-seat ampitheater and a 25-acre amusement park, complete with aquatic rides, on a portion of this federal parkland. (In exchange, the developers were to provide financial assistance for park maintenance and rehabilita-

RISING SEAS IN NEW YORK CITY

Development is not, of course, the only issue facing the city's coast. The greatest long-term concern is the worldwide problem of sea level rise. Government scientists now believe that carbon dioxide gases produced from the burning of coal, oil, and other fossil fuels are trapping heat that would otherwise escape from the earth's surface and lower atmosphere. This is often referred to as the "greenhouse effect."

If global warming trends continue, water levels along the Atlantic coast may climb between two and seven feet during the next century. Even over the next several decades, the impacts of sea level rise on New York's beaches, airports, and other low-lying areas could be considerable. Rising waters along New York's coast could be particularly troubling in areas of the city already susceptible to erosion and flooding problems, such as Coney Island and the Rockaways. Sea level rise may also pose substantial risks to the region's drinking water supply.

Of course, the heavy responsibility for reversing global warming trends remains primarily in the hands of decision makers in both national and international circles. But state and local governments can do a lot by themselves to encourage energy conservation and slow global warming. And they will also have to take sea level rise into account in coastal land-use decisions of the 1990s.

tion.) Park Service officials finally shelved the controversial leasing arrangement in the wake of intense public opposition.

And then there's Staten Island. Development pressures in recent years have centered less on the coast than on inland parcels. Property owners and the state Department of Environmental Conservation have been locked in dispute over the department's reclassification of 1,300 acres as freshwater wetlands. The redesignation requires landowners to secure state permits prior to construction; DEC can deny such permits where development would damage the wetlands themselves. The controversy pits the economic interests of development companies and some private landowners against such environmental interests as flood and stormwater control, and the protection of fish and wildlife habitat.

The Law: The Coast

The United States Congress has never been wild about land use planning. It has repeatedly rejected legislative attempts to involve the federal government in what it considers to be state and local prerogatives. But in the heyday of congressional activism on the environment in the early 1970s, a federal coastal land use statute was swept into place. The little-known Coastal Zone Management Act of 1972 offers funding to states that draft plans for the management and protection of their coastlines. The statute does not demand that coastal states adopt such plans. But to cash in on the federal largess, state coastal projects must, among other things, provide for the preservation of beaches, wetlands, barrier islands, and other natural resources, protect flood- and erosion-prone areas and assure recreational opportunities and public access to the coast.[42] The Commerce Department's National Oceanic and Atmospheric Administration administers this federal program.

With both coastal protection and federal dollars on its mind, New York passed enabling legislation that created a state coastal zone management program in the early 1980s. At the heart of New York's federally approved plan are 44 broad policy objectives. Their purpose: to incorporate the federal coastal protection priorities into state decision-making. New York State invited local governments to share the federal proceeds. To secure this funding, municipalities were required to adopt their own coastal plans. New York City did just that, bolstering its coastal protection authority, at least in theory.[43]

Laws and regulations concerning coastal development turn up everywhere. Among other statutes, the Clean Water Act mandates that persons seeking to discharge dredged or fill material into navigable waterways secure a permit from the Army Corps of Engineers. And under the 1899 Rivers and Harbors Appropriations Act, permits from the Corps are also required for the placement of structures in navigable waterways. Both federal and state law direct the preparation of environmental impact statements in connection with major actions, including certain coastal development projects that could significantly impact environmental quality. And under the state law, agencies taking actions that have been the subject of an environmental impact statement must explicitly find that adverse environmental impacts will be minimized to the maximum practicable extent.[44]

Other little-known state laws come in handy when talk turns to coastal resources. One empowers the state to identify and map areas prone to coastal erosion and to require permits for construction-related activities in these regions. Two additional statutes set up a mapping and permitting process to help shield saltwater and freshwater wetlands from development pressures.[45]

At the local level, it is the city's Zoning Resolution that calls the shots. This maze of rules is particularly complex, and even an incomplete summary could stretch on for pages. Briefly, the Zoning Resolution, in conjunction with the Building Code, regulates the construction and use of buildings throughout the city. Among many other things, it divides the city into districts, specifying permissible uses within those areas. And it imposes further controls on the size and, in some cases, the design of buildings. While a variety of zoning provisions covers properties located on the waterfront, the resolution itself contains no comprehensive set of requirements designed specifically to regulate development along the city's coast.

Government Action: The Coast

Will the real New York City coastal program please stand up? Is it a set of coastal priorities that protects natural resources, promotes water-dependent uses, increases waterfront public access and recreation, and advances only those development proposals that are compatible with these objectives? Or is it a policy that leaves the door open for waterside commercial and residential development on almost any scale, as typified by the colossal Trump City proposal?

Until now, the city has been sending mixed signals. For one thing, it has embraced the state's coastal zone management program. It has created within the city Planning Department a local unit to apply these coastal priorities to waterfront land use decisions. This operation is sometimes referred to as the Waterfront Revitalization Program.

At least on the margins, the city's presence is making a difference. Of the more than 1,100 projects that have come before the department between 1984 and 1988, roughly 30 percent were later modified to conform with coastal zone management objectives, planning officials report. For example, a proposal to locate a Manhattan community hospital on the Harlem River was altered to include a public walkway along the water's edge, following review by the Planning Department. And on at least one occasion, the department's intervention actually derailed a proposed waterfront project, which it found inconsistent with the coastal plan. That's why you won't find a dockside commercial car wash built on pilings in Brooklyn's Mill Basin.

But hold the champagne. New York's coastal zone management plan, as currently structured, has considerable limitations. Governmental reviews of proposed projects are hampered by the plan's broad and sometimes conflicting policy objectives. As two of the city's leading environmental lawyers, Stephen L. Kass and Michael B. Gerrard, have noted, the plan provides no formal mechanism for balancing competing coastal management goals (i.e., economic development versus protection of natural resources). Further, since the coastal management policies and standards are so numerous, projects can seldom move forward without compromising at least one plan objective or another.

Despite these shortcomings, the city Planning Department's Waterfront Revitalization unit has at least attempted to balance competing coastal interests. It would be hard to say the same for the city's Public Development Corporation (PDC). This quasi-governmental agency was tapped in 1985 to spearhead development projects along the coast. Its five-borough development wish list includes, among other things, proposals for platforming over segments of the city riverfront and effectively selling off portions of city parks. Economic factors, not protection of coastal resources, seem to dominate the PDC agenda. Indicative of the agency's apparent mind-set is its 52-page, 1986 blueprint that mentions the role of the city's federally approved coastal zone management program not even once.[46]

The city's approach then remains somewhat schizophrenic. And by failing to speak with a single voice, it may be letting private interests set the city's agenda along the waterfront. As Robert A. Caro, an award-winning author and longtime commentator on urban planning in New York, has noted, "For the city to just react to a developer's plan is to abdicate its moral authority as government."

FIGURE 2.4: NEW YORK CITY PUBLIC DEVELOPMENT CORPORATION'S PROJECT WISH LIST (1986)

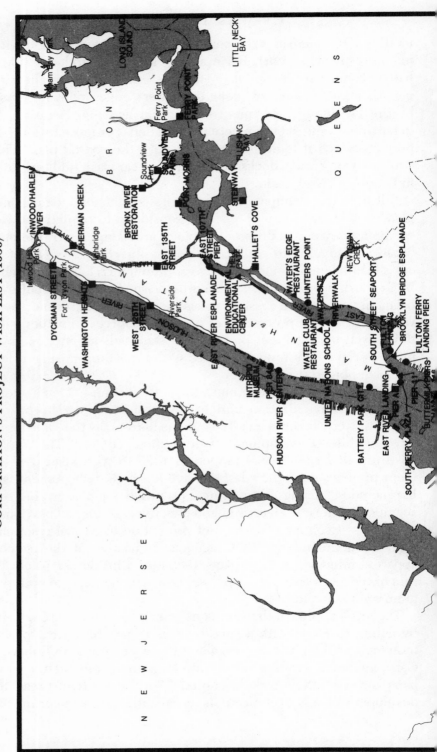

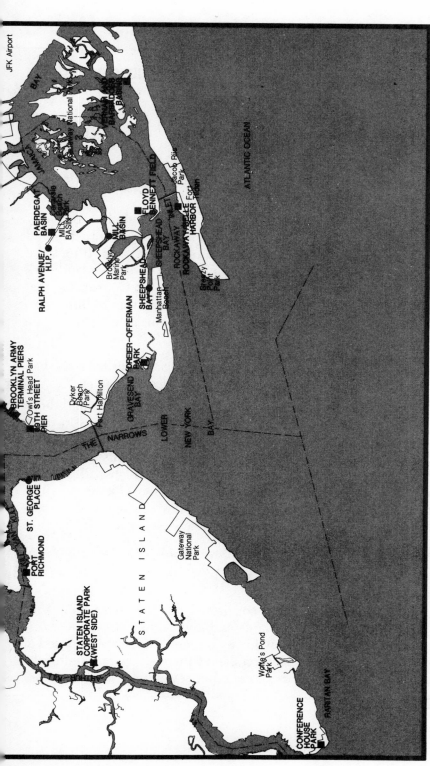

New York City Public Development Corporation

● **Current Projects**   ▲ **Completed Projects**
■ **Sites under Study**   —— **Completed Public Improvements**

JFK Airport

JAMAICA BAY

Gateway National

PAERDEGAT BASIN

RALPH AVENUE/ H.I.P.

Mill Basin Park

MILL BASIN

Brooklyn Marine Park

BROOKLYN ARMY TERMINAL PIERS

Owl's Head Park

39TH STREET PIER

Dyker Beach Park

Fort Hamilton

GRAVESEND BAY

DREER-OFFERMAN PARK

SHEEPSHEAD BAY

Manhattan Beach

SHEEPSHEAD BAY

FLOYD BENNETT FIELD

Jacob Rije Park

Fort Tilden

ROCKAWAY INLET

ROCKAWAY/TRIPLE HARBOR

Breezy Point Park

ATLANTIC OCEAN

THE NARROWS

LOWER NEW YORK BAY

ST. GEORGE PLACE

PORT RICHMOND

STATEN ISLAND

STATEN ISLAND CORPORATE PARK (WEST SIDE)

Gateway National Park

Wolfe's Pond Park

CONFERENCE HOUSE PARK

RARITAN BAY

OCEAN DUMPING: QUESTIONS AND ANSWERS

Q. *What is ocean dumping?*

A. Ocean dumping is the disposal of treated or untreated waste products in coastal waters.

Q. *Is garbage being dumped into the waters off New York City's coast?*

A. No, at least not legally. Contrary to popular belief, ordinary household and commercial trash has not been lawfully dumped in New York's coastal waters for more than 50 years. It took the United States Supreme Court to close the door. In a landmark case, *New Jersey v. City of New York,* the High Court ruled that garbage cast into the ocean by New York City was washing onto New Jersey's shore and creating a public nuisance. It ordered a halt to the practice, which took effect in 1934.

Q. *What about the garbage and medical wastes that have washed ashore in New York and New Jersey in recent summers?*

A. Government officials do not yet know the full story behind these incidents, which left highly publicized beach closings in their wake. It is likely that several different sources were responsible. Medical wastes and other debris flushed out of city sewers and refuse unintentionally spilled at the city's Fresh Kills landfill and trash transfer stations were among the most likely contributors.[47] Compounding the problem is raw sewage washed into regional waters, particularly after heavy rainfalls.

Q. *Do government officials permit any wastes to be jettisoned into New York's coastal waters?*

A. Yes—three types:

1. *Sewage sludge.* This consists of residues from New York City sewage treatment plants and eight other sewage authorities in the New York–New Jersey region. Sludge is composed of water, sewage solids, bacteria, heavy metals, organic chemicals, and other contaminants screened at sewage works. Nearly 3.9

million wet tons of sludge are being dumped by New York City alone into ocean waters every year. The eight other sewage authorities add about 4.6 million tons to the annual total.

2. *Dredge materials*. To keep marine navigational channels and docking areas clear, mud, silt, and sediments are regularly scooped from the bottom of New York Harbor. These dredge spoils frequently contain heavy metals, organic compounds, and other contaminants that have accumulated along the sea floor. More than 8 million cubic yards a year of these wastes are barged to off-shore disposal grounds.

3. *Construction debris*. Concrete, excavation dirt, rubble, and rock have been dumped, albeit in decreasing amounts, off New York's coast since 1940. No such material was disposed of in local waters during 1986 or 1987. But one company, the New Jersey–based Port Liberte Partners, holds an EPA permit to dump 400,000 cubic yards of excavation dirt; through April 1989, however, the firm had disposed of only 17,000 cubic yards of this material.[48]

Q. *Where is ocean dumping taking place?*

A. *Sewage sludge* is released from barges at a deepwater (6,000 to 9,000 feet) site located roughly 138 miles southeast of New York Harbor. This site is often referred to as the 106-mile ocean dump site (as measured from New Jersey's shore). For years, sewage sludge had been dumped at a 12 mile offshore site; that practice ended in 1987 by order of the U.S. EPA.

Dredge materials from the harbor floor are dumped at a relatively shallow (50 to 80 feet) site, about five miles south of New York's Rockaway Peninsula.

And *construction debris* is tossed to sea at a site about seven miles off New Jersey's shore, just south of Sandy Hook.

Q. *What are the impacts of ocean dumping?*

A. Closure of shellfish beds due to high bacteria counts. Elevated levels of heavy metals and toxic organic compounds in bottom sediments and marine organisms. Smaller fish catches. Repro-

FIGURE 2.5: NEW YORK OCEAN DUMP SITES

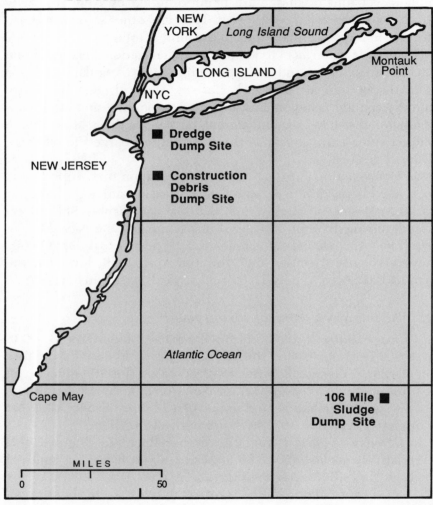

NRDC

ductive difficulties and increased incidents of fin rot and other fish diseases. These and other adverse ecological effects have been observed in and around the so-called 12-Mile Sewage Sludge Dump Site, which was used for the disposal of municipal sludge until EPA ordered a halt to the practice there in 1987. At that time, EPA concluded that ocean dumping of municipal sludge was most likely the primary cause of such problems in this portion of New York's coastal waters and was a contributing factor to the overall degradation of the larger New York Bight.

Dredge disposal has presented similar problems. One has been the closure of shellfish beds due to high bacteria counts and elevated levels of heavy metals and organic compounds. Unique to dredge disposal is the physical burial of bottom-dwelling marine organisms following each dumping. In total, sludge and dredge deposits have contributed as much as 50 percent or more of some heavy metals and PCBs to the New York Bight.

There is no consensus on the impacts of dumping at the recently established 106-mile ocean dump site. Some officials maintain that the environmental impacts are minimal due to dilution and dispersal of sludge in deeper waters, more active currents, and lower fish and shellfish populations. But fishermen claim that recent declines in lobster harvesting and increased incidents of burn-spot disease (a shell-destroying malady) are linked to sludge dumping at this deepwater location.[49] A definitive analysis of the situation has not yet been completed, although the impacts of sludge dumping at the old 12-mile site certainly suggest troubling problems.

Little information is available concerning the ocean dumping of construction-related wastes.

Q. *What law governs ocean dumping?*

A. The Marine Protection, Research and Sanctuaries Act of 1972, often referred to as the Ocean Dumping Act, authorizes EPA to regulate the disposal of all types of materials into ocean waters and to strictly limit the dumping of any material that will adversely affect human health, welfare, or the marine environment. The act and its implementing rules also require EPA to designate approvable sites for ocean dumping. And the statute also charges the Army Corps of Engineers, subject to EPA review, with administering the issuance of permits for dredging and limiting the disposal of contaminated dredge spoils.

In the wake of the summer of 1988 beach washups, Congress amended the Ocean Dumping Act. The 1988 amendments are the latest congressional attempt to end sludge dumping once and for all in the coastal waters off New York. Under the statute, all municipalities must cease dumping operations by December 1991 or face escalating fines.[50] (New York City and several communities north of the city and in New Jersey are the only municipalities in the nation that continue the ocean dumping of sewage sludge.)

New York City officials, faced with some difficult choices, have agreed (in what is probably not the last word on this issue) that the city will be out of the ocean by 1992. Composting, landfilling, and perhaps even incineration are the land-based alternatives that city officials will be pushing into the limelight in the mid-1990s. Regardless of the final disposal route selected, the most important next step is to reduce the heavy metal concentrations that now make sludge so unwelcome in all quarters.

PART 3

Air Pollution

They point to Mexico City. Or maybe Athens, Greece. Now there's a place with *real* air quality problems, some New York apologists argue. Of course, they have a point. In Mexico City, 3 million largely uncontrolled motor vehicles pour as much as 4 million tons of contaminants into the air every year; during the winter smog season, morning visibility can fall to 500 yards. Things have gotten so bad that a few Mexico City residents keep emergency gas masks or oxygen canisters in their homes or offices. Across the globe in Athens, pollution eating away at the Parthenon and the other monuments on the Acropolis has prompted government officials to move some statues and other priceless antiquities indoors to protect them from further airborne attack.

No, the air in New York City is not nearly that awful. But stop an out-of-towner along the streets of New York on a summer day, or ask an elderly asthma or emphysema sufferer, and you're likely to get another perspective. Or contact a homeowner or apartment dweller who lives along a bus street or heavily trafficked thoroughfare, or near a dry cleaner, sewage treatment plant, or municipal incinerator; you may get an earful about just how air pollution is affecting the health and quality of life of many New Yorkers every day in 1990.

There are cross-cutting reports from the air pollution battlefront. Even the most hardened pollution street fighter has to acknowledge significant reductions in some air contaminants over the past 20 years. The widespread shift to low sulfur oil as the major home heating and utility fuel has brought dramatic declines in sulfur dioxide. Federal curbs on gasoline lead levels have largely cleansed city air of that toxic contaminant. And even carbon monoxide, the colorless, odorless poison that helped build New York City's reputation as a pollution capital, is no longer found at concentrations five times higher than national health standards. (Although carbon monoxide levels still surpass the federal health limit, you can thank automobile pollution control devices for what reductions have been achieved.) In some important respects, New York City air is measurably cleaner than it was two decades ago.

Still, these successes have been accompanied by much slower progress in other areas, by the discovery of additional pockets of resistance, and by new trends that, if not corrected, could pave the way for declines in air quality. Ozone smog, its brownish-yellowish haze hanging over the city during summer months, remains a serious problem. The growing proportion of new trucks that are diesel-powered (which are joining over 5,000 diesel buses already on city thoroughfares) means even more street-level discharges of sooty particulate matter. Increasing motor vehicle travel throughout the region threatens to wipe out much of the progress that has already been made in reducing all vehicular pollutants. And scores of hospital incinerators and other toxic dischargers only now being identified are sending hazardous fumes into city air every day.

But what is quite possibly most surprising about air quality in New York City is how much cleaner our air could be. Buses that run on alternative fuels like natural gas hold the promise of dramatic reductions in street-level emissions. Pollution controls on such toxic generators as hospital incinerators and dry cleaners are technologically available today. And implementation of regionwide measures can cut back even on the seemingly intractable problem of ozone smog. Although it won't be easy, New York City's still-troubling air quality problems can be solved.

BACKGROUND

For most New Yorkers, air pollution is their principal exposure to environmental contaminants. This includes everything from cigarette smoke, indoor pollution, and occupational exposures to motor vehicle discharges and localized sources of toxic pollution. New Yorkers drink perhaps two quarts of water a day. In that same period, they breathe approximately 35 pounds of air. On the whole, city residents still enjoy high-quality water. But if they didn't, they could always choose to purchase bottled water. New Yorkers have no choice as to the air they breathe.

Our air is a mixture of gases—about 78 percent nitrogen and 21 percent oxygen, along with trace amounts of carbon dioxide and such other compounds as methane, carbon monoxide, hydrocarbons, and particulates. The earth's atmosphere is several hundred miles high. But the density of its air thins out quickly with altitude. Most of its total air mass is concentrated in the lowest atmospheric layer that surrounds the earth's crust. It is a thin band of tropospheric air no more than about ten miles high that receives the brunt of the world's pollution loadings.[1]

Synergistic is a word that environmental scientists like to throw around. In air pollution circles, it is used to describe the phenomenon in which the combined exposure to a potpourri of contaminants multiplies the adverse impact that the individual substances themselves would have. For example, risks from asbestos or radon skyrocket if the person exposed is also a cigarette smoker. This is an extremely important concept for New Yorkers. While most scientific studies of air pollutants track exposure to a single chemical, city residents are inhaling dozens of different compounds every day. The impacts of such multiple exposures are worrisome, even though a precise quantification of risks is impossible. (Remember, too, that the impacts of a particular pollutant also depend on such factors as age, occupation, and health of persons exposed.)

Environmental regulators, especially those concerned about air pollution, often find themselves along the frontiers of scientific uncertainty. For many pollutants, monitoring and exposure data are scarce. Definitive epidemiological proof is difficult to come by.

TABLE 3.1 TEN MAJOR AIR POLLUTION SOURCES
IN NEW YORK CITY

Source	Number	Typical Pollutants
Automobiles	Approximately 1.9 million registered	Carbon monoxide, hydrocarbons, nitrogen oxides
Diesel-powered trucks and buses	Approximately 12,400 registered trucks and 5,000 buses	Particulates, sulfur oxides
Electric and steam generating plants	Thirteen	Sulfur dioxide, nitrogen oxides, particulates
Home and apartment house heating systems	Approximately 830,000	Sulfur dioxide, nitrogen oxides, particulates
Incinerators	Three existing municipal, 103 hospital, and approximately 2,200 apartment house	Particulates, sulfur dioxide, nitrogen oxides, heavy metals, assorted toxins
Industrial facilities	Fourteen sewage treatment plants, unknown number of other sources	Hydrocarbons, assorted toxins
Commercial establishments	Approximately 2,500 gasoline stations, approximately 2,000 dry cleaners; unknown number of other sources	Hydrocarbons, perchloroethylene
Airports	Two major facilities	Carbon monoxide, hydrocarbons, nitrogen oxides, particulates
Indoor pollutants	Widespread	Carbon monoxide, nitrogen oxides, formaldehyde
Out-of-state generators	Unknown	Ozone, particulates, sulfates

NRDC

TOBACCO ROAD

WARNING: Your authors have determined that cigarette smoking is crazy.

We are not alone in this conclusion. Former U.S. surgeon general C. Everett Koop has bluntly labeled cigarette smoking "the chief, single avoidable cause of death in our society and the most important public health issue of our time."

Even by New York City standards, tobacco smoke is a big-time air pollutant. It contains more than 4,700 chemical compounds, including arsenic, benzene, carbon monoxide, carcinogenic tars, formaldehyde, nicotine, nitrogen oxides, sulfur dioxide, and vinyl chloride, to name a few. It can be inhaled directly into the lungs both by smokers themselves and by family members, co-workers, and others who are exposed to its noxious fumes. And it is discharged in large amounts—every year roughly 467,000 tons of tobacco are smoked indoors throughout the United States.[3]

Enormous is the right way to describe the toll tobacco smoking takes on public health. It is the leading cause of lung cancer, responsible for more than 85 percent of nationwide deaths from that disease. It more than doubles the risk of heart attack. And it makes those who smoke two to three times more likely than nonsmokers to suffer strokes, the nation's third leading cause of death (after heart attacks and cancer).

There's more. According to the federal Centers for Disease Control, tobacco use boosts the risk of cancer of the mouth, esophagus, larynx, stomach, cervix, bladder, kidney, and pancreas. Smokers have a total cancer death rate two times greater than nonsmokers (four times greater for heavy smokers). And smokers are more prone to chronic lung diseases like asthma, bronchitis, and emphysema, as well as to such other ailments as stomach ulcers, migraine headaches, and pneumonia. Every year, 390,000 Americans—including perhaps as many as 11,000 New York City residents—

die from diseases directly related to smoking.[4]

You need not actually smoke cigarettes (or other tobacco products) to be harmed by them. Simply breathing air containing tobacco smoke (sometimes referred to as "passive" or "involuntary" smoking) increases your risks. A single smoker in a home can double the amount of particulate pollutants inhaled by nonsmoking householders. In homes where smoking occurs, such smoke is usually the leading source of airborne mutagens (agents that can induce changes in the body's cells). The U.S. Environmental Protection Agency estimates that nationwide as many as 5,000 nonsmokers may die every year just from lung cancer caused by involuntary smoking.[5] Some scientists believe that the impacts of passive smoking on

the heart of exposed nonsmokers is of even greater magnitude.

There is some good news on the smoking front—smoking rates, on the whole, are declining. The percentage of Americans who puff cigarettes has been dropping by half a percent a year since the first smoking warnings were issued by the surgeon general's office 25 years ago. In 1964, 40 percent of the U.S. adult population smoked. In 1987, the national number was down to 28.8 percent. In New York, a 1987 state Health Department estimate put the citywide figure for cigarette smokers at 19.5 percent.[6]

But the overall decline in cigarette smoking has not spread evenly across all segments of the population. Smoking rates nationally for black males continue to exceed those for white males by a

With the exception of a few of the most studied contaminants, decision-makers must sometimes act based on limited data, or risk exposing large populations to suspected hazardous substances for years.

Under these circumstances, Congress has concluded that the wiser course is to err on the side of caution. As a federal appeals court in Washington, D.C., said in a famous 1976 case involving an industry challenge to regulations that would phase down levels of toxic lead in gasoline: "Regulatory action may be taken before the threatened harm occurs; indeed the very existence of such precautionary legislation would seem to demand that regulatory action precede, and optimally prevent, the perceived threat."[2]

Despite such sentiments (which polls show are widely shared

sizable margin (39 percent to 30.5 percent in 1987). (New York City experts believe the ratio here is comparable.) Black smokers are an important segment of the cigarette market and they have the illnesses to prove it. Since 1950, the rate of lung cancer death among black men has risen three times faster than that for white men. And one federal analysis concluded that the contribution of smoking to mortality rates was 20 percent greater for blacks than for whites. (Smoking rates for women in high school and college and for blue-collar workers are also higher than average.)

In the last few years, New York lawmakers have begun clamping down on public smoking. They are responding to the growing political militancy of nonsmokers and to continuing revelations from medical experts like the New York Lung Association as to the health risks from passive smoking. New York City's statute, which took effect in 1988, restricts smoking in virtually all enclosed public places from taxis to indoor sports arenas. Under its more controversial provisions (which seem to be working quite nicely, thank you), restaurants with more than 50 seats have set aside half their seating for nonsmoking diners and offices with open work areas and more than 15 employees are required to provide a smoke-free environment to employees who request it. New York State's Clean Indoor Air Act, which took effect in January 1990, extends the smoke-free guarantee to all work places, regardless of size.[7]

among New Yorkers), the debate over scientific uncertainties continues to creep into almost every major air pollution dispute.

Which New York City neighborhoods face the most severe air quality problem? Although air pollution levels across the city are not uniform, the question of which community is breathing the most polluted air is complex. For one thing, the answer differs depending on the pollutant. Ozone smog affects all parts of the city to a more or less equal degree. Most other pollutants have a more localized impact. Carbon monoxide and diesel particulates, generated almost exclusively by motor vehicles, are worse in Manhattan and along heavily trafficked thoroughfares in the other boroughs. Staten Island residents receive the lion's share of the largely unquantified industrial emissions from nearby New Jersey.

New York City Dept. of Transportation

SLOW GOING. *Despite major cutbacks in emissions from autos rolling off production lines, motor vehicles remain the dominant source of air pollution in New York City. Carbon monoxide levels in our air are lower than 20 years ago but still violate national health standards. Partly responsible for the continuing problems are growing vehicular travel throughout the region and weaknesses in the state's annual auto emissions inspection program.*

But while outlying neighborhoods in Brooklyn and Queens may have lower levels for many pollutants, residents in those communities should not rest easy if they happen to live across the street from a localized pollution source like a gasoline station or bus stop. And regardless of what community they call home, city residents may well be inhaling high levels of pollution caused by cigarette smoke and other indoor contaminants. Air pollution, in short, is an equal opportunity contaminant.

MOTOR VEHICLES

Introduction

Looking for the number-one source of air pollution in New York City? Forget (for the moment) about incinerators, power plants, or industrial dischargers. Just look out the window to the nearest street or avenue. Motor vehicles are far and away the dominant source of airborne contaminants. Every year in New York City, automobiles, trucks, and buses send hundreds of thousands of tons of pollution right into city air. No other pollution source even comes close.

. You might be wondering about catalytic converters and the other antipollution doodads that have been installed on every automobile sold in the United States since the mid-1970s. If they are cutting auto emissions, why is there still a problem? One reason is that New Yorkers hold on to their cars for quite a while. Roughly one-third of the passenger cars registered in the New York metropolitan area are nine or more years old, according to the state Department of Motor Vehicles. Pollution control devices often don't perform at top efficiency for this length of time. So the big pollution reductions from autos rolling off the production line fall off sharply in the latter years of a vehicle's life.

A second reason why motor vehicles still chalk up big pollution numbers in New York City is that there's more to the motor vehicle fleet than autos alone. Roughly 200,000 trucks, for example, are registered (and are revving their engines) in the five boroughs. Tens of thousands more pour into the city to make deliveries and conduct their daily commerce. But federal pollution reduction mandates have let heavy-duty vehicles off the hook for years. Pollution limits that really mean something don't even take effect for diesel-powered trucks until 1994. Other polluting vehicle engines such as those in construction equipment and motorcycles have also largely escaped effective control. Then there are the buses. They constitute Exhibit A in the case against poorly controlled pollution sources.

Finally, let's not forget the number of miles traveled by all motor vehicles in New York City and throughout the region. Care to guess which way that number has been going? Between 1970 and 1987, total miles traveled in the area have shot up by more than 20 percent. The situation, obvious throughout the city, is especially severe in Manhattan. Over 760,000 cars squeezed into the central business district (south of 60th Street) on a typical workday in 1988. This is over 100,000 more daily entries than in 1980.[8]

Motor Vehicles: The Law

From a legal standpoint, if you're talking motor vehicle pollution, the heavy artillery is the federal Clean Air Act. Shaped in 1970 and amended in 1977, this granddaddy of environmental laws seems on its way to a further revision in 1990. Among the most important provisions in the existing law is a requirement that all automobiles sold in the United States achieve 90 percent reductions in emissions of carbon monoxide and ozone-forming hydrocarbons (75 percent for nitrogen oxides), as compared to 1970 models.[9]

The law also directed EPA to reduce sharply levels of poisonous lead in gasoline. And it required states like New York that recorded unhealthy levels of air pollution to adopt clean air plans (providing for cleanup programs like vehicle emission inspections and traffic reduction measures). If motor vehicle pollution in New York City is less oppressive in 1990 than it was in 1970 (and to some extent it is), credit goes in part to environmental friends like former senator Edmund Muskie (Maine) and former congressman Paul Rogers (Florida), two early architects of the nation's clean air law.

With the Clean Air Act's motor vehicle provisions alone stretching to more than 20 pages (along with hundreds of pages of more detailed federal rules), state and local legislative and regulatory actions have played supporting roles. The state's federally required clean air plan has been the focal point for Albany's clean air activities (chief among them being the annual automobile emissions inspection program, described in more detail later). At the municipal level, two of the few meaty provisions are city laws that restrict motor vehicle idling to no more than three minutes and that prohibit visible tailpipe emissions after a vehicle has been traveling for more than 90 yards.[10]

DIESEL-POWERED VEHICLES

New Yorkers know a problem when they smell one. So, few people who have caught a whiff of an accelerating diesel bus or truck will be shocked to hear that emissions from diesel-powered vehicles are a big pollution issue in New York City. This town is the nation's diesel bus capital. Every day, well over 5,000 public and privately owned buses roll across the city's streets and highways. (Chicago, with the nation's second-largest fleet, has about 2,500.)

Trucks, tens of thousands of them, also lumber through town on a daily basis. They range in size from light-duty vans to double-trailered behemoths. An increasing proportion of them are diesel-powered. A significant number of trucks are city-owned; the Sanitation Department alone has about 1,800 diesel-powered garbage collection trucks. According to government figures, trucks log roughly 10 percent of total vehicle miles traveled in the five boroughs and are responsible for even more of the total vehicle pollution discharges. They are the second largest source, behind buses, of diesel emissions in New York City.

The diesel engine, as we now know it, was patented in 1892 by Rudolf Diesel (no surprise there). In the two-stroke diesel engine, fuel is injected directly into the cylinder and detonated by heat and pressure, without the use of spark plugs. The diesel engine's major advantage is fuel efficiency and economy. It runs on a cheaper, less-refined type of petroleum than gasoline. For such advantages, New York City's air pays a high price.

The diesel engine's major problem is particulate pollution. Diesel-powered vehicles emit 30 to 70 times more particulate matter than similarly sized gasoline-powered engines. These particles are released from the incomplete combustion of diesel fuels. They differ from particles emitted from many stationary pollution sources because they are overwhelmingly smaller and inhalable. (More than 90 percent are less than 2.5 microns in diameter; the thickness of a human hair is about 100 microns.) Although no recent official estimates are available, a 1985 analysis that relied on city figures indicated that perhaps 3,000 tons a year of diesel particulates are spewed out by motor vehicles in New York City.[11]

The danger from diesel particulates is compounded because they are discharged directly at breathing level. The emissions, although most serious in Manhattan, are not confined to that borough. Diesel trouble spots can be found on main traffic arteries, bus depots, loading zones, and other locations in all five boroughs.

The health case against diesel particulates is a strong one. The tiny size of these fine particles allows them to elude the body's natural filtering mechanisms in the nose, throat, and lungs. Particles reaching the lower portions of the lung may remain there for weeks or even years. Diesel particulates are generally composed of a carbon core on which dozens of different toxic compounds may be absorbed. Among them are hazardous organic substances such as benzopyrenes and other known and suspected carcinogens. The U.S. Environmental Protection Agency has projected that in 1995 diesel exposure alone will account for between 2 percent and 8 percent of the lung cancer risk faced by nonsmokers. [12]

Diesel particulates are linked to medical problems other than lung cancer. They inhibit the lung's ability to clear particles, bacteria, and virus from the respiratory tract. They contribute to or aggravate chronic, obstructive lung diseases (asthma, bronchitis, and emphysema) and pose added risks for patients with cardiovascular problems or influenza. Finally, certain compounds that adhere to diesel particulates have been shown to be mutagenic (capable of causing genetic alterations) in animal tests. [13]

How large is the diesel particulate problem in New York? Once again, limited monitoring data make generalizations difficult. In fact, it was only in 1988 that the first street-level monitor was installed in a heavily trafficked Manhattan street canyon, where one would expect to find the highest readings. (Next time you stroll down the east side of Madison Avenue, between Forty-seventh and Forty-eighth streets, you may notice this contraption mounted on a tripod and protected by a chain link fence.) Sure enough, the first year of data from the Madison Avenue monitor revealed that pollution levels there are exceeding the newly revised national health standard for particulate matter. The 1988 annual average, according to state figures, was more than 56 micrograms per cubic meter of air; the standard is 50 micrograms. On some days, readings at this bus-busy location were more than two times the national health limit.

Of course, diesel particulate levels at less heavily traveled sites are lower, and there are many other sources of particulate matter throughout New York City. But as the National Academy of Sciences has observed: "The small size and chemical properties of the [diesel] particulates may cause problems disproportionate to their total contribution to the mass of suspended particles in the air." [14]

Particulates are the diesel's main problem, but not its only one.

Scott McKiernan

STILL DIRTY AFTER ALL THESE YEARS. *With more than 5,000 Transit Authority and private express buses, like those pictured above, booming down city streets every day, New York City is the nation's diesel bus capital. These buses and a growing number of diesel-powered trucks pour 3,000 tons a year of toxic particulates into the city's air. Cleaner-burning fuels like natural gas may hold the promise of eliminating these noxious diesel fumes.*

Nitrogen dioxide is being found in higher concentrations in New York City these days, and emissions from diesel engines are one of the likely sources. Diesel fuel can also contain up to seven times more sulfur than gasoline. As a result, diesels produce far greater sulfur dioxide and sulfate emissions than do gas-powered vehicles.

Emissions from diesel engines also degrade urban visibility. The small size of diesel particles allows them to remain suspended in the air for long periods. And their black, sooty carbon cores give them a light extinction efficiency three or four times greater than other fine particles. According to the Environmental Protection Agency, how far one can see in cities like New York may fall 12 percent by 1995 from levels in the mid-1970s—primarily due to increased diesel emissions. And that is a projected annual average; impacts on individual days are expected to be even worse.[15]

Diesel engines are also potent producers of dirt. Many hundreds of tons of diesel particulates are deposited on buildings, structures, and clothing every year in New York City. Diesel particulates are blacker and oiler than typical airborne particles, making them stickier and more easily smeared. The millions of dollars spent here every year on exterior building cleaning—for which diesel emissions deserve a fair share of the blame—is one of many hidden economic costs of air pollution in New York City.

As if all this weren't enough, you'll be smelling even more diesel exhaust in the 1990s. This will result from a major shift quietly taking place in the composition of the motor vehicle fleet. The small proportion of cars that are diesel-powered is not expected to change much. But light-duty trucks that have run on gasoline (the smaller ones, commonly used as urban delivery vehicles) are frequently being replaced these days with vehicles that operate on diesel fuel. (The fleet of yellow *New York Post* newspaper trucks you may have seen around town is just one example.) As a result, federal officials project an 11 percent increase in urban area diesel pollution between 1985 and 1995. And that assumes that federal emissions limits for new buses and trucks take effect as scheduled in the early 1990s.

GASOLINE-POWERED VEHICLES

Step right up for a few facts you're not likely to hear from "the good Olds boys" (or any other car dealers, for that matter). Every year, the average automobile (traveling 10,000 miles) emits about 650 pounds of carbon monoxide, 105 pounds of hydrocarbons, 50 pounds of nitrogen oxides, and 12 pounds of particulates into the air. Motor vehicles are also the single largest contributor of airborne toxic contaminants in urban areas like New York. And they are a significant emitter of carbon dioxide, a major contributor to global warming. A single tank of gasoline produces about 300 to 400 pounds of CO_2 when burned. Over 1,900,000 passenger cars are registered in New York City alone, to say nothing of the hundreds of thousands of vehicles that stream into the five boroughs every day from other parts of the region. We are talking here about a fair amount of pollution.[16]

With environmental data often uncollected or inaccessible, insiders sometimes turn for hard-to-get numbers to environmental impact statements. Usually prepared by outside consultants, these documents (required by law for projects that may have significant environmental impacts) are often chock-full of handy facts. So it is that the environmental impact statement prepared for the city's plan to put 400 additional taxis on city streets (a plan that has been shelved for the time being) contains the only available projection for total 1989 motor vehicle discharges in New York City for three major contaminants: carbon monoxide—almost 473,000 tons; hydrocarbons—more than 49,000 tons; and nitrogen oxides—over 36,000 tons.[17] Since such estimates vary, these numbers should be viewed as giving only a general sense of the scope of the problem.

A brief discussion of pollution trends for several vehicle-related pollutants follows:

Carbon Monoxide. Nobody would have picked carbon monoxide as the pollutant that set off what was probably the city's fiercest environmental battle of the 1970s. After all, carbon monoxide is a gas that is colorless, tasteless, and odorless. But levels of this poison, pouring out of automobile tailpipes, were being monitored in the air at levels up to five times the national health standard. Nevertheless, New York City's 1973 clean air plan (which included such controversial strategies as tolling the East River bridges and restrictions on midtown parking) had been quietly resting on agency bookshelves, in suspended animation. (Carbon monoxide interferes with the oxygen-carrying capacity of the blood, causing reduced awareness, dizziness, headache, and fatigue in heavy traffic conditions. It is most dangerous to cardiovascular patients since it further stresses their body's already limited oxygen delivery system.)

When two creative lawyers, Ross Sandler and David Schoenbrod, filed suit in 1975 on behalf of environmental groups to enforce the city's plan, the stage was set for a dramatic confrontation. Mayor Abraham Beame howled on the front pages that implementation would turn New York City into a "ghost town" and directed city lawyers to mount a full-scale defense. But in several landmark rulings, the federal courts upheld the environmental plaintiffs. "[C]itizen groups are not to be treated as nuisances or trouble-makers but rather as welcomed participants in the vindication of environmental interests," said the U.S. Court of Appeals for the Second Circuit on one such occasion.[18]

The U.S. Supreme Court refused to hear the city's appeal and, from a legal standpoint, that was that. But with widespread opposition to bridge tolls among city motorists, Congress (which was amending the Clean Air Act at the time) was coaxed into authorizing then governor Carey to delete the toll strategy, which he did. Meanwhile, the parties to the litigation were working out compromise strategies to establish exclusive bus lanes on city streets and strengthen enforcement of existing traffic laws, among other things. By the end of 1977, with Mayor Beame's term in office drawing to a close, the conflict had been diffused, even if all the carbon monoxide had not.

Over the last 15 years, carbon monoxide readings in New York City have declined, although the levels are not yet out of the danger zone. Curbside monitors have been moved around town like deck chairs on the *Titanic* and improved traffic control around monitoring sites can skew the results, so accurate comparisons of maximum carbon monoxide exposure levels are difficult to make. But according to state data, the peak eight-hour concentration in 1988 was 13.9 parts per million (ppm) (recorded at 59th Street between Third and Lexington avenues). This is over the 9 ppm national standard, but considerably less than the city's 1973 peak eight-hour reading of nearly 50 ppm. Still, projections for 1989 carbon monoxide levels at 12 heavily traveled intersections in Manhattan forecast violations of the standard at 10 of them. Real progress has been made in reducing CO levels, but there still is a way to go.

Ozone. Just defining the ozone issue can be tricky. You've probably heard about the hole in the ozone layer. That refers to the ozone shield in the upper atmosphere, about 18 miles or so above the earth's surface. This protective layer, which filters out ultraviolet radiation, is being punctured by chlorofluorocarbons and other man-made chemicals. Scientists believe that further deterioration of the ozone layer will lead to increased incidence of skin cancer and eye problems in humans and will harm forests, crops, and animal life around the world.

Those troubling developments, however, have nothing to do with the ozone problem that is the focus of this section. We are concerned here about ground-level ozone, a widespread air pollutant that is itself responsible for serious, adverse health and environmental impacts. (To keep the two ozone problems straight, just remember that there's too much of it down here, and not enough of it up there.)

Ground-level ozone is a gaseous pollutant formed in the atmosphere when hydrocarbons and nitrogen oxides mix in the presence of sunlight. Ozone is the primary constituent of photochemical smog, the yellowish-brown haze that often envelops New York City and other urban areas, especially during warmer weather. Ozone at levels frequently experienced here during those periods can cause breathing difficulties, chest tightness or pain, and inflammation of the lungs even in otherwise healthy persons. Chronic exposure to ozone at such levels can result in increased susceptibility to respiratory infection, impaired breathing capacity, and permanent damage to lung tissues.

Where do the hydrocarbons necessary for ozone formation come from? In the 1970s, hydrocarbon sources were more or less evenly divided between motor vehicles on the one hand and everything from gasoline stations and industrial solvents to consumer paints and hair sprays on the other. More recently, the contribution of motor vehicles has slightly decreased (presumably as a result of auto pollution controls). In the New York region, motor vehicles accounted for about 45 percent of the area's total hydrocarbon emissions in 1987 (the most recent year for which data are available) and is presumably continuing on a slow downward curve.[19]

You can't talk to state or city air pollution officials for more than five minutes before they remind you that ozone is a regional problem. They have a point. As opposed to carbon monoxide (which dissipates quickly) or diesel particulates (which hit their highest levels close to heavily trafficked roadways), a fair amount of the ozone you breathe in New York City probably had its origins in New Jersey, Pennsylvania, or points even farther west. Similarly, the impacts of hydrocarbon discharges in New York City are often felt most severely on Long Island, in Westchester, or in southern Connecticut, which has recorded some of the highest ozone levels in the Northeast.

Regardless of how many jurisdictions contribute to New York City's ozone problem, none can deny that it remains a serious problem. Violations of the national air quality standard were recorded at all six of the New York metropolitan area's ozone monitors in 1988. (The highest one-hour average of 0.206 parts per million, close to twice the 0.12 standard, was measured at the Queens College site.) After some improvements in the 1970s, ozone levels in New York more or less stabilized in the 1980s. In fact, during the hot summer of 1988, the number of ozone exceedances was the highest of the decade. That year, ozone violations were recorded at one of the New York metropolitan area's monitoring stations on 28 separate days.

Lead. Barry Commoner kicked up some dust among environmental activists with a provocative *New Yorker* article back in 1987. In this retrospective look at environmental gains and losses since 1970, the year of the first Earth Day, Commoner suggested that environmental programs have succeeded only where government has prohibited the production of pollutants (pollution prevention) and have generally failed whenever government has sought to regulate the allowable level of discharges (pollution control).[20]

Regardless of the degree to which you embrace his overall conclusion (and environmentalists are still debating that one), you've got to admit that when government flat out bans an environmental poison, progress can be dramatic. Take toxic lead in gasoline. In 1973, new federal rules commanded gasoline marketers to begin reducing levels of lead in their product. When they started, more than 220,000 tons of the heavy metal were being discharged into the nation's air from motor vehicles. But by 1987, when the long-delayed but triumphant lead-in-gas phasedown regulations had taken full effect, gasoline lead discharges had been slashed by more than 95 percent.

In New York City, the story of airborne lead could put a smile on everyone's face. Levels have fallen off drastically over the last 15 years, a direct result of the phasedown of lead in gasoline. The highest quarterly average recorded at state-operated lead-monitoring sites in 1988 was 0.21 micrograms per cubic meter of air (at the Greenpoint, Brooklyn, sampler); this is well below the national health standard of 1.5 micrograms. Air lead levels in the New York region tumbled by 90 percent or more between 1973 and 1988, essentially eliminating (at least for now) lead in the air as a public health problem in New York.

Toxics. Aren't all air pollutants toxic? If you define "toxic" literally as a poison, the answer is yes. In the language of the Clean Air Act, there's a slightly more complicated answer. Since the early 1970s, EPA has devoted major resources to regulating six widespread pollutants (carbon monoxide, ozone, nitrogen oxides, sulfur dioxide, particulates, and lead). These are referred to not as "toxic" but as "criteria" pollutants (since the act required EPA to draft documents establishing criteria or standards for acceptable levels of these pollutants in the air). Most other pollutants are usually lumped into the "toxic" category. They range from synthetic organic chemicals (like benzene) and inorganics (such as chlorine) to fibers (like asbestos) and metals (such as mercury).

Calculating the health risks from air toxics couldn't be tougher. There are major gaps in data on actual emissions. There are differing models for estimating levels of toxics in the air. And there is the ever-present, but unquantified, synergistic effect from multiple pollutants. The U.S. Environmental Protection Agency has taken a first cut; in 1989, the agency estimated that up to 2,700 cancer deaths nationwide are caused each year by public exposure to a selected number of air toxics. EPA also reported that the lifetime excess cancer risks from exposure to the toxic "urban soup" of chemicals may range from 1 in 10,000 to 1 in 1,000.[21] (Over the years, EPA has often used a risk level of 1 in 1 million as a guidepost for triggering regulatory action.)

The major contributor to toxic air pollution in New York City may surprise you. Although scientists caution that there is still much uncertainty about air toxins, EPA's 1989 assessment concluded that motor vehicles were the single largest contributor to cancer risks from exposure to air toxics. Motor vehicles, said EPA, are responsible for 55 percent of the total cancer incidence from air contaminants, five times greater than from any other air pollution source. In the EPA survey, motor vehicle particulates accounted for more than 76 percent of the benzene, 63 percent of the directly emitted formaldehyde, and 77 percent of all polycyclic organic matter found in urban air.[22]

Government Action

Citizens looking for relief from air pollution in New York City have come to learn a new meaning for the word *patience*. In 1965, the federal Motor Vehicle Air Pollution Control Act gave the secretary of health, education and welfare (there was no EPA in those days) the authority to prescribe emissions standards for new motor vehicles. The president then was Johnson. The mayor was Wagner. In the decades that followed, federal, state, and city officials have all had opportunities to reduce motor vehicle contaminants. They have taken some; they have missed others. Despite noteworthy gains in some areas, government officials will have to step up the pace of their actions if they don't want to hear New Yorkers complaining about motor vehicle pollution problems in the twenty-first century.

The federal motor vehicle pollution control program has been the flagship of the national clean air campaign. To achieve automotive emissions standards required by the 1970 Clean Air Act amendments, vehicle manufacturers have had to install catalytic converters and other sophisticated pollution control devices in cars coming off assembly lines. (In a catalytic converter, exhaust gases are circulated over a metal catalyst and mixed with air, changing carbon monoxide to carbon dioxide, and hydrocarbons to carbon dioxide and water.) Vehicle emissions standards have trimmed average carbon monoxide and hydrocarbon pollution levels for new automobiles by 75 percent or more (although inadequate maintenance of pollution control equipment, increasing volatility of gasoline, and growing travel throughout the region have eaten away at these reductions). Despite the remaining problems, reduced automotive emissions brought about under this program represent one of the federal government's more significant environmental accomplishments over the last two decades.

Federal actions to combat escaping pollutants from buses and trucks have been less impressive. Heavy-duty vehicles went essentially uncontrolled for years. Only in 1988 were manufacturers of diesel-powered vehicles required to get into the act. Tight emissions limits are not scheduled to kick in until 1991 (for urban buses) and 1994 (for other buses and all trucks). It's always possible that EPA will fold under industry pressure and postpone even these long-awaited deadlines still further.

After being deadlocked for much of the 1980s, Congress may finally in 1990 give the Clean Air Act an urgently needed facelift. Among key things to watch for on the motor vehicle front: (1) do the amendments mandate two new rounds of emissions standards, or leave a second round at the discretion of EPA (a prescription for further delay)?; (2) do the amendments include improved warrantee and recall provisions to insure that pollution control devices perform as promised?; (3) do the amendments include a strong program for alternative-fuel vehicles?; and (4) do the amendments provide increased leverage for regional solutions to regional air pollution problems?

Decidedly mixed is the state's record in helping to control New York City's top air pollution problem. It takes a well-tuned, well-maintained car to maximize pollution reductions. For that reason, many states have established automobile inspection programs to check emissions systems and require tune-ups or repairs if necessary. New York State's program has been run out of more than 4,000 gasoline stations in the nine-county metropolitan area. It has been a major disappointment.

The state's emissions inspection program has been problem-plagued since its inception in 1981. First, the stringency of its emissions test was so low that even vehicles with high levels of discharges were passing. Then, in the mid-1980s, it was discovered that nobody could account for perhaps as many as 500,000 inspection stickers. This is not what you call effective quality control. Repeated audits by the U.S. Environmental Protection Agency identified continuing problems with the program's effectiveness, despite state efforts to clean up the mess. If the program were revamped and run out of state-operated centralized inspection stations (15 states already use such a system), it could, according to preliminary state calculations, bag over 200,000 tons in annual carbon monoxide reductions and perhaps 14,000 tons of hydrocarbon cuts.[23] The Cuomo administration could do little that would be more important for clean air in metropolitan New York than to unveil a centralized emissions inspection program.

Meanwhile, give the state and DEC commissioner Thomas C. Jorling credit for making some smaller, but still important, moves to clean the city's air of motor vehicle–related contaminants. It now requires large- and mid-size gasoline service stations in the nine-county metropolitan area to capture ozone-producing gasoline fumes that otherwise escape into the air during vehicle refueling. (You may have noticed the new "vapor recovery" nozzles at local gas stations.)[24] It also has helped organize northeastern states in a regional program requiring the oil industry to reduce the evaporative content of gasoline sold here. Finally, New York and its northeastern sister states have agreed to impose tighter new car emissions standards, which are now applicable only in California.

Do the governor and state legislature have the courage to take bold action? If they do, and they are looking for the one measure that can help check seemingly incessant increases in motor vehicle travel and generate funds to meet urgent public transportation needs, they might raise the state tax on gasoline. This environmentally sound, fiscally prudent approach has been taken in one way or another by all but 4 of the 50 states in the 1980s, according to the Highway Users Federation in Washington, D.C. New York State gasoline levies have not gone up since 1972. With gas prices relatively stable now, there will never be a better time to act.

At the local level, New York City has little to brag about for its efforts to reduce motor vehicle pollution. The showpiece in a spartan cupboard is its alternative fuels program. Presently, the program consists of two demonstration projects, testing in buses that are powered by alternative (nondiesel) engines. In one test, six methanol-fueled buses donated to the city by the General Motors Corporation (in settlement of an unrelated clean air case) have been whisking Queens bus riders around since 1988. The Triboro Coach Corporation is operating these vehicles. In the second demonstration, the Brooklyn Union Gas Company has equipped two buses with engines that run on natural gas. The buses, operated by the Command Bus Company, have been serving Brooklyn riders for the last two years.

Alternative fuels hold considerable promise for reducing motor vehicle pollution in New York City, although their exact role is still unclear. Right now, such projects seem to make most sense for urban buses and trucks. Preliminary emissions test results from the two bus demonstration projects show sharply decreased particulates from both fuels, although higher emissions of several other pollutants. (Meanwhile, the New York City Transit Authority is testing an add-on pollution control device, called a trap oxidizer, which it hopes will be able to control diesel fumes without having to switch to a cleaner, alternative fuel.)

A comprehensive study of the two fuels by INFORM, a New York environmental think tank, concludes both fuels have air quality benefits, but that natural gas is preferable to methanol from a pollution reduction perspective.[25] Also waiting in the wings are electrically powered vehicles (which would emit no pollution at the point of use) and further down the line perhaps the ideal fuel—hydrogen (which may be produced from renewable solar power). The Dinkins administration seems likely to accelerate the city's alternative fuel-testing program. May the best fuel win.

City officials have made little progress on the tough task of reducing the amount of motor vehicle traffic in the city and in the region. But, unless they do, increased emissions from growing travel will erode pollution reductions from other strategies. Placing some restrictions on auto entries into the most congested parts of Manhattan entices traffic planners. Transit Chief Robert Kiley and some city transportation officials recognize that the city's economy is dependent not on private autos but on public mobility. They believe that subduing traffic congestion and reclaiming street space for pedestrians and transit are necessary to protect Manhattan's attractiveness for tourists and shoppers. But when some politicians hear these ideas, they run for the hills. And the city's Department of Environmental Protection has historically provided little leadership, seemingly happy to putt the issue off the fairway.

Car pooling and ride sharing are less controversial. But even these sensible traffic reduction measures have not received much attention from city (or state) officials in recent years. Presumably, they'd be dusted off again in the next gasoline crisis; there is little to lose and much to gain by stepping up activity in this area now.

IF YOU'RE THINKING OF BREATHING ON ...
STATEN ISLAND

Welcome to Staten Island, where, more than in any other borough of New York City, residents will tell you they worry about air pollution in their community. Although the evidence behind such concerns is not yet definitive, what is already known has some people worried:

• Staten Island has for years recorded the highest respiratory cancer death rates in the city;
• Staten Island residents continue to complain about odors in the air they breathe. (In 1988, more than 1,000 air pollution complaints from Staten Islanders were received by the Interstate Sanitation Commission, many more complaints than were filed that year by any other county within the tri-state agency's jurisdiction.)[26]

Environmental scientists like Dr. John Oppenheimer at the College of Staten Island believe that some of the problems can be traced to industrial facilities in nearby New Jersey, which is separated from Staten Island only by a narrow waterway called the Arthur Kill. At over 120 industrial and commercial facilities in three neighboring New Jersey counties, more than 10,000 tons a year of volatile organic substances alone shoot or seep into the air. Often

Increased reliance on bicycle transportation might not dramatically reduce pollution levels in New York City. But improving opportunities for the city's estimated 65,000 bicycle commuters—a matter that has been largely ignored by city officials since a bike-lane experiment was terminated in the early 1980s—could induce additional ridership and have important symbolic value.

The whole area of transportation and air quality planning could receive a shot of adrenaline in the next few years. Under federal law, the city is once again going to be revising its clean air plan. In anticipation, the city Transportation Department has commissioned a comprehensive study on traffic congestion and pollution reduction alternatives. Longtime observers off the city scene get the distinct impression they've seen all this before. But as with they are carried by prevailing winds westerly into Staten Island. Environmentalists also identify the Fresh Kills landfill and ever-increasing motor vehicle traffic as possible sources of odors and complaints.

Government officials are less certain about the nature of the air quality problem on Staten Island. They acknowledge that, like the rest of the region, Staten Island has an ozone problem, especially in warmer weather, and that higher particulate levels have been recorded on Staten Island than in many other parts of the city. But when it comes to airborne toxics, they say they just don't know how much worse, if at all, conditions are on Staten Island. They need more information.

The U.S. Environmental Protection Agency, and the states of New York State and New Jersey, are finally committing some resources to help close the data gap. Spurred by Staten Island borough president Guy Molinari, they have stationed 13 toxic air pollution monitors at strategic points throughout the borough and in New Jersey. It's part of their impressive-sounding Urban Air Toxics Assessment Project. Preliminary data from one monitor reveal trace levels of formaldehyde in concentrations more than double the recommended state guideline, as well as detectable levels of benzene and other toxic contaminants.[27] The full study is expected to be completed later this year.

---

other aspects of the air quality problem, you'd be surprised what a strong push from the federal government and a new cooperative spirit between regional officials might accomplish.

COMMERCIAL, INDUSTRIAL, AND RESIDENTIAL SOURCES

Power Plants and Space Heating

It was one of the earlier victories on the city's environmental battlefront. (If you are under 30, you probably haven't heard the account.) And it remains one of the sweetest—city and state officials joining forces to impose and retain some of the nation's toughest environmental restrictions on fuel burning.

It is not astonishing that utility power plants and home and apartment house heating systems have for years been a huge source of air pollution in the city. The burning of coal, oil, and to a lesser extent natural gas (so-called fossil fuels) creates unwanted by-products. The main troublemakers are sulfur dioxides, nitrogen oxides, and particulates. When you look at New York City's 13 electric and steam generating plants and at the heating systems in more than 800,000 homes and apartment buildings, you get some idea of how much burning goes on in the nation's most densely populated city. [28]

The pollutants from fossil fuel combustion can be quite nasty. Sulfur dioxide is a caustic, colorless gas that can constrict the bronchial air passages and trigger asthma attacks in persons with that disease. It also interacts with other compounds in the atmosphere to produce sulfates and sulfuric acids, which return to earth as acid rain. Nitrogen dioxide, a yellowish-brown gas emitted from all fuel-burning processes, primarily attacks the respiratory system. Exposures to this pollutant can increase the incidence of acute respiratory illness, increase susceptibility to infection, and damage lung tissue.

The good news started rolling in when the City Council took action in the 1960s to end the burning of high-sulfur fuel. (In those years, you could find a new layer of soot and grime on your window sill every day; much of it came from fuel oil burning.) In 1966, a new city law set the first limits (1 percent) on the sulfur content of heavy fuel oil burned in New York City. (Before that, the sulfur content was typically 2 percent or higher.) The law also required an upgrading of fuel-burning equipment to reduce particulate discharges. Then, in 1971, a new air pollution control code ratcheted the content of sulfur in fuel down to 0.30 percent, one of the tightest standards anywhere. [29] Congressman Ted Weiss, then a young city councilman, was one of the reformers behind the ground-breaking legislation.

As a result of these actions, there has been a dramatic fall-off in emissions from power plants and home and apartment house heating systems in New York City. Once again, comparing emissions in a data-starved field is no cinch. But according to available city figures, Con Edison's sulfur dioxide emissions, which were 325,000

Eric A. Goldstein

CLEANER. *Sulfur dioxide levels in New York City air have declined dramatically over the last two decades. This is a result of city laws that sharply limit sulfur content in fuel oil burned in home and apartment house boilers and in utility power plants, like Con Edison's huge Ravenswood facility, pictured above.*

tons in 1966, had fallen to 35,000 tons by 1977. In 1989, Con Edison reports, total emissions of sulfur dioxide from its facilities were less than 20,000 tons. Annual citywide emissions of sulfur dioxide from all sources were approximately 58,000 tons in 1985, the most recent year for which full data are available.[30] (There have also been reductions in utility emissions of nitrogen oxides, although not nearly as dramatic. And since low-sulfur fuels are also cleaner burning, particulate discharges have declined as well.)

Concentrations of sulfur dioxide in the city's air have followed the reduction of this pollutant in fuel. In the 1960s and early 1970s, sulfur dioxide levels regularly crested into the unhealthy range.

But these days, all city monitors show compliance with the national health standard of 0.03 parts per million. (Still, the highest 1988 SO_2 readings in the state—0.024 ppm—were recorded at the Mabel Dean Bacon Vocational High School Annex monitor on Manhattan's Lower East Side. Peak sulfur dioxide levels in the city have held steady at about two-thirds of the national standard throughout the 1980s.)

It's a different story for nitrogen dioxide. At one Manhattan location, concentrations of this pollutant have been hovering just below the national health standard in recent years. In 1988, for example, the monitors at P.S. 59 in upper midtown recorded the city's highest readings of 0.046 parts per million. (The national standard is 0.05 parts per million.) Officials, who have not paid much attention to this situation, are not quite sure why this is occurring. (Levels of nitrates and sulfates in city air have also been creeping up in recent years; interstate transport is apparently the major source.)

In the early 1980s, Con Edison sought to throw the sulfur dioxide success story into reverse. The utility requested permission to burn 1 percent sulfur coal (instead of 0.30 percent sulfur oil) at two of its main power facilities (Ravenswood, Queens, and Arthur Kill, Staten Island). The move, an effort to cut utility costs at the expense of air quality, would have increased sulfur dioxide emissions in the city by more than 66,000 tons a year. Opposition to this ill-considered plan was led by City Council president Carol Bellamy. And in one of his most significant decisions, Henry Williams, state DEC commissioner at the time, conditioned approval of coal burning on the requirement that Con Ed install pollution-capturing scrubbers. The utility reluctantly abandoned its coal-burning scheme shortly thereafter. Should the proposal resurface again in the 1990s, the environmental troops will be prepared.

There is one other caveat. If your apartment or office sits just above or adjacent to a neighbor's smokestack, the fact that overall, citywide emissions from home and apartment house boilers have declined provides little comfort. You don't need us to tell you that if you are in the path of a black plume from a recently fired-up boiler, you have a problem. But this site-specific pollution threat has not yet been the subject of intensive study or investigation. It is an issue that warrants some nosing around by government scientists.

TABLE 3.2 HOSPITAL INCINERATORS PERMITTED IN NEW YORK CITY

Brooklyn

Community Hospital of Brooklyn
Coney Island Hospital
Cumberland Hospital
Kingsbrook Jewish Medical
  Center (2)
Kings County Hospital Center (4)
Lutheran Medical Center
Maimonides Medical Center (3)
Methodist Hospital (2)
St. Mary's Hospital (2)
Victory Memorial Hospital
Woodhull Medical Center

Bronx

Albert Einstein College Hospital (3)
Beth Abraham Hospital
Bronx–Lebanon Hospital Center
Bronx Municipal Hospital Center (2)
Bronx Psychiatric Center
Daughters of Jacob Center
Lincoln Medical and Mental Health
  Center
Montefiore Medical Center
New York Zoological Society Animal
  Health Center (2)
North Central Bronx Hospital
Pelham Bay General Hospital
St. Barnabas Hospital (2)
Union Hospital of the Bronx
Veterans Administration Hospital
Westchester Square Hospital
Workman's Circle Home for the
  Aged (2)

Manhattan

Bellevue Hospital Center (2)
Beth Israel Medical Center (3)
Cabrini Medical Center
Coler Memorial Hospital
Columbia Presbyterian Medical
  Center (3)
Doctors Hospital
Goldwater Memorial Hospital (2)

Gracie Square Hospital
Harlem Hospital Center (2)
Hospital for Special Surgery
Joint Diseases North General
  Hospital (3)
Lenox Hill Hospital (3)
Manhattan Eye, Ear and Throat
  Hospital
Memorial Hospital for Cancer and
  Allied Diseases (3)
Metropolitan Hospital Center
Mount Sinai Hospital
New York Eye and Ear Infirmary
New York Hospital, Cornell Medical
  Center (2)
New York Psychiatric Institute
New York University Medical
  Center (4)
Rockefeller University Hospital
St. Luke's Roosevelt Hospital (3)
St. Rose's Home
St. Vincent's Hospital and Medical
  Center of New York
Terence Cardinal Cooke Health Care
  Center
Veterans Administration Medical
  Center

Queens

Booth Memorial Medical Center
Flushing Hospital and Medical Center
Jamaica Hospital (2)
Long Island Nursing Home
Mary Immaculate Hospital
Peninsula Hospital Center (2)
Physicians Hospital
Queens Hospital Center
St. John's Queens Hospital (2)
St. John's University (2)

Staten Island

Doctors' Hospital of Staten Island
Staten Island Developmental Center
Staten Island Hospital

Source: New York City Department of Environmental Protection; New York State Department of Environmental Conservation (12/89)
  * Includes incinerators of regular garbage, infectious, and pathological wastes; a number of these permitted incinerators are not presently in use.

Incinerators

Garbage-burning incinerators have been smoking in New York City for decades. Over the last 20 years, however, the winds of reform have been slowly blowing away this pollution menace, at least for the time being. City Air Code reforms in the 1960s and early 1970s are largely responsible. Faced with upgrading requirements, all but three of the antiquated Sanitation Department's municipal incinerators have shut down operations. (At one time there were 22.) The drop in on-site apartment house incinerators is also occurring, although too slowly—from a peak of about 17,000 to roughly 2,200 today. The long-overdue total phaseout of these apartment house burners is, under a new City Council law, to occur by 1993.

Although we describe the incineration problem in detail in part 1, we reiterate two points here. First, the nearly 100 existing hospital incinerators in New York City remain a pesky source of localized pollution. They share the problems of poor combustion efficiency and inadequate pollution controls. As the amount of disposable medical waste continues to increase, hospitals and other medical institutions will likely seek to burn greater amounts of such trash. Unless these incinerators are shut down (replaced by an intensive medical waste reduction program and by a regional, state-of-the-art medical trash incinerator), protests by citizens in surrounding neighborhoods will, with some merit, probably pick up further steam.

Finally, there's the question of huge new garbage incinerators, which the Sanitation Department still hopes to construct—at least one in each borough. The air quality impacts of such a move could be quite sizable, as the figures included in our part 1 discussion of the proposed Brooklyn Navy Yard facility reveal. There is not yet a definitive study, however, that attempts to calculate the combined citywide pollution loadings from five or more of such giant burners. While government attention on the proposed Navy Yard incinerator is now riveted on the issue of ash disposal, there is likely to be renewed concern about air quality impacts from proposals for a series of additional incinerators. Government officials who want to pursue this route will have to convince a skeptical citizenry that more incinerators won't mean a reversal of 20 years of hard-fought pollution gains.

Petroleum Marketing

Gasoline fumes, wherever they are—in auto engines, in vehicle fuel tanks, in service stations, or in huge storage tanks like the ones you see on the New Jersey Turnpike—are desperately trying to escape into the atmosphere. They often succeed. This makes the petroleum industry a significant contributor to ozone-producing emissions and air toxins in the New York region. The nine-county New York metropolitan area has more than 3,000 service stations from which gasoline fumes are emitted. And five of the top ten individual sources of volatile organic compound emissions in New York City are bulk petroleum storage terminals, according to raw data obtained from the state Department of Environmental Conservation. (The others include two manufacturers with large-scale painting and finishing operations, an asphalt plant, a piano manufacturer, and a large commercial baker.)[31]

Environmental officials have begun cracking down on some petroleum emissions, but others continue to elude control. In the last several years, the state has finally imposed new rules requiring large- and mid-size service stations to capture gas fumes otherwise emitted during vehicle refueling. It has also ordered petroleum refiners to decrease the evaporative content of gasoline sold here. Both moves should help to cut thousands of tons of ozone emissions on an annual basis. But regulations aimed at reducing emissions during transfers of gasoline from gas tanker trucks to service station tanks do not appear to be enforced aggressively by either city or state environmental agencies.

Consumer Solvents and Paints

Some people would be delighted if reducing air pollution in New York meant that only businesses and industries would have to do the reducing. When it comes to ozone smog, however, things are not so simple. On the motor vehicle side, attaining clean air here will require better-tuned, better-maintained vehicles and more car pooling and ride sharing. In the home, it will mean cutting back the use of highly evaporative paints and solvents, and getting used to changes in such familiar products as hair sprays, insecticides, and air fresheners. Somewhat surprisingly, state data indicate that consumer use of paints, solvents, and other products with evaporative organic chemicals is the third leading source of ozone emissions here, behind motor vehicles and industrial processes.

There has already been a stirring of government action in this area. In 1989, state rules took effect that trim the level of evaporative compounds in most oil-based paints and some other coatings sold in New York. And the state is also committed to reducing the volatility of five consumer products—hair care products, antiperspirants, air fresheners, insect repellents, and pet sprays. Some of these products will have to be reformulated, or might even be taken off the market before the calendar hits the year 2000.

Sewage Treatment Plants

You have to concede that it's slightly embarrassing. The region's sewage treatment plants, so important to the cleansing of rivers, bays, and beaches, are polluting the air in New York City. According to the U.S. Environmental Protection Agency, treatment works in the region may be emitting more than 10,000 tons a year of ozone-generating volatile organic compounds as well as airborne toxics. Although little monitoring has been done in New York City, among the organic compounds measured in ambient air immediately downwind of sewage plant aeration tanks at one Rhode Island facility were xylenes, benzene, trichloroethylene, and 1,1,1-trichloroethane.[32]

Environmental scientists know that the problem doesn't start at the treatment plants themselves. The real sinners are the industrial firms who use the sewers to dispose of their unwanted by-products, without first adequately pretreating their chemical wastes. At municipal sewage plants, significant portions of these organic chemicals turn into vapors that escape into the environment. The majority of such emissions in the region come, EPA believes, from treatment plants in New Jersey. (Average industrial discharges to sewage works are almost five times greater in the Garden State than in New York.)

As is often the case, it is occupational exposures that may be most risky. But sewage plant workers are not the only ones who are being affected. Residents on Manhattan's Upper West Side in the vicinity

SURPRISE GUESTS. *Some of the sources of air pollution in New York City are quite unexpected. They include hospitals (incinerator particulates), dry cleaners (perchloroethylene), airports (carbon monoxide and other fuel-related contaminants), sewage treatment plants, and certain consumer products (volatile organic compounds). Aggressive government action could significantly reduce emissions from all five categories.*

of the North River treatment facility report that odors from the plant have been causing breathing difficulties, headaches, and smarting eyes, on and off, for years. A new citizen's group, West Harlem Environmental Action, has helped focus attention on their concerns. They are still waiting for an adequate government response.

We don't need Houdini to make such emissions disappear. The technology is already well known. According to EPA, sewage treatment plants may be "one of the most feasible and more easily achievable categories" left for cutting smog-producing emissions. Despite the reluctance of cost-conscious municipalities on both sides of the Hudson, it's likely that controls on sewage plant vapors will have to be included in both states' new clean air plans. But to battle-weary neighbors of some of the facilities, promises of odor reduction still seem like a mirage.

Airports

More than just planes are going into the sky at John F. Kennedy and La Guardia airports. These two Queens County facilities, essential to New York City's mobility and economic health, are also an overlooked source of air pollution. For one thing, they are big-time auto trip generators. Over 5,000 cars an hour jam the access roads to JFK during peak periods, according to the Port Authority of New York and New Jersey. A second problem: nearly 1,000 landings and takeoffs a day at La Guardia and over 800 a day at Kennedy send exhaust and fumes into the air. Finally, there are the escaping vapors from aviation fuel storage and transfer. In what is probably incomplete accounting, one government estimate suggests that 5,500 tons of hydrocarbons, 5,000 tons of nitrogen oxides, and 19,000 tons of carbon monoxide, along with unknown quantities of particulates and air toxics are generated at these two travel hubs every year from aircraft emissions alone.[33]

Like air travel itself, air pollution from these facilities will climb in the 1990s. The Port Authority expects passenger volumes, already at record highs at both airports, to soar—25 percent at La Guardia and 45 percent at Kennedy—by the year 2000. This will mean more flights, more fuel, more auto trips, and more emissions. (And let's not forget anticipated growth at New Jersey's Newark and New York Orange County's Stewart airports, both of which add to regional pollution loads.)

Government officials on the airport scene are scurrying to handle the expected passenger surge. Pollution abatement is not high on their list. The folks at the Port Authority, which operates the two giant airstrips, are planning a multiyear, $3.2 billion redevelopment project for Kennedy. The project even has a name: "JFK 2000." What it doesn't have (and what La Guardia doesn't have, either) is a direct airport rail link. So the number of auto trips, and resulting auto emissions, will continue to climb well into the next century. And although there are some controls on emissions from aviation fuel storage and transfer, the federal government has done little to control airplane engine emissions. Expect the airports to grow as pollution pockets over the next decade.

Other Industrial and Commercial Sources

When it comes to pollution from industrial or commercial facilities in New York, the big problem is from small sources. According to a recent study by the American Lung Association of New York State, which reviewed reported emissions from major facilities only, New York City ranked well down the list of total toxic emissions.[34] But before breathing a sigh of relief, remember that chronic low-level emissions from small-scale operations such as dry cleaners, service stations, and printers (and, of course, motor vehicles) contribute most of our background toxins. Their cumulative health effects on New Yorkers are still unquantified.

Still, there are a handful of major industrial sources. One of the larger ones is the Brooklyn-based Ulano Corporation. This is a plastics coating factory in Boerum Hill, Brooklyn. It is a major discharger of volatile organic compounds (over 700 tons in one recent year, according to unverified state records). And, at least as of 1987, it was the city's largest source of the toxic pollutant toluene. For years, the U.S. Environmental Protection Agency and the state have been seeking through administrative action to get the Ulano

Corporation to install state-of-the-art controls. EPA reports that some progress is being made. But as of late 1989, the legal proceedings, and the pollution, were continuing.

More typical of the city's toxic generators are its estimated 4,000 dry cleaning establishments. The contaminant of concern here is perchloroethylene, a pungent cleaning solvent that is a probable human carcinogen. According to EPA, typical dry cleaning machines emit about eight pounds of perchloroethylene for every 100 pounds of clothes cleaned. Well-designed facilities with carbon adsorber controls could reduce such discharges to a tiny fraction of that amount.[35] So far, there is little information available as to how many dry cleaners in the city have installed state-of-the-art controls. Nor is there at this point any legal directive requiring them to do so.

The overall government response to toxic air pollution is still in its infancy. The federal Clean Air Act, at least until now, has not been especially effective in getting industrial facilities to install state-of-the-art control technology. For the most part, the state's efforts have also been anemic. New York's limits for concentrations of airborne toxins are in most cases too lenient and in all cases only recommended (and unenforceable) guidelines. Although the State Department of Environmental Conservation is planning to upgrade its toxics program in the 1990s, as things now stand, there are not even criteria for when the department should conduct stack monitoring of toxic emissions from industrial or commercial facilities. Nobody seems to remember, for example, the last time a hospital incinerator in New York City was checked by stack probes.

The best news about industrial toxics is that the issue seems finally to be getting more attention. Right-to-know laws are arming citizens with information on toxic emitters in their communities. The Community Environmental Health Center at Hunter College, for instance, used such data to prepare a comprehensive assessment of toxic neighbors in Greenpoint/Williamsburg, Brooklyn.[36] And a few state and federal environmental officials now seem more interested in getting to the bottom of the air toxics story. One can only hope that such activities will lead to a better identification of the most significant toxic threats and to stepped-up efforts to cleanse the city's air.

TABLE 3.3 SOME COMMON SOURCES OF INDOOR AIR POLLUTION

Source	Selected Pollutants
Cigarettes and other tobacco products	Respirable particulate matter, carbon monoxide, benzene, carcinogenic tars, aldehydes and many others
Insecticides	Organophosphates
Pesticides	Petroleum distillate
Mothballs	Paradichlorobenzene
Kerosene and gas heaters	Carbon monoxide, sulfate aerosols
Gas ranges	Nitrogen oxides
Fireplaces	Particulates, creosote
Copying machines	Ozone
Dry cleaning fluids	Perchloroethylene
Paints	Volatile organic compounds
Some carpet adhesive backings	4-phenylcyclohexene
Chipboard	Formaldehyde
Some foam insulation	Formaldehyde
Cosmetic talcum powder	Silica fibers
Uranium-bearing rocks and soil	Radon
Air "fresheners"	Paradichlorobenzene

NRDC

Indoor Pollutants

Most New Yorkers spend up to 90 percent of their time indoors. So you'd think that the quality of indoor air would have been a big issue with government pollution fighters and the environmental community. But that has not been the case. With the federal Clean Air Act all but ignoring the topic, it has only been during the last several years that indoor air pollution has begun to attract the attention it deserves.

There's little disagreement now as to the importance of the matter. The U.S. Environmental Protection Agency reports that

A WORD ABOUT THE WORLD'S MOST IMPORTANT ENVIRONMENTAL ISSUE

With the risk of superpower nuclear war receding, perhaps the most significant environmental issue for the world today is global warming. There is now a scientific consensus that emissions of carbon dioxide, methane, and other gases into the atmosphere are bringing about climatic change. This is occurring as such gases absorb heat that radiates from the earth's surface and reflect some of that heat back to earth, warming the planet. If present trends continue, scientists calculate, temperatures could increase by three to eight degrees Fahrenheit by the middle of the next century. The debate is no longer whether the so-called greenhouse effect is real, but only over magnitude and timing of the forthcoming shifts. As the National Academy of Sciences has reported, "We are already irrevocably committed to major global change in the years ahead."[42]

Is global warming an issue for New York City? Consider the possible impacts projected in one recent study by the Urban Institute:

• The average number of days here with temperatures above 95°F could increase from less than 3 to 22, boosting summertime electricity demand by 10 to 20 percent.

indoor concentrations of some air pollutants may be two to five times outdoor levels. A comparative risk assessment that the agency completed several years ago ranked indoor air pollution as among the country's most significant environmental health problems.[37] And, as if medical concerns were not enough, indoor air pollution is a significant economic drain—problems with "sick buildings" and lost productivity are costing employers tens of millions of dollars a year.

The sources of indoor air pollution are many. In some homes, it is cigarette smoke, which may be the highest indoor source of toxic benzene and particulate air pollution. In others, it is kerosene heaters and gas ranges, which could be large contributors of carbon monoxide or nitrogen oxides. Carpet adhesives, chipboard, and certain foam insulation—all three and more may be sources of

- Higher temperatures could cause ocean expansion and glacier melting, raising sea levels by between 1.5 and 6.5 feet—leading to increased sewer backups, basement flooding in low-lying areas, and vast new spending programs for dikes, levees, and other construction projects to protect private property and city infrastructure.
- Losses of city water ranging from 10 to 24 percent could result from increased evaporation at surface reservoirs and from lower snow accumulations in the watershed; at the same time, water demand could be increased by perhaps 5 percent to accommodate increased cooling of large buildings and more intensive lawn watering. (Drinking water shortfalls throughout the region could squeeze New York City's supply even further.)[43]

It is beyond the scope of this book to discuss in any detail the problem of global warming or remedial actions that must be taken to slow down atmospheric temperature shifts. But it is enough to note that state and even local governments can play an influential role in this campaign. Since fossil fuel burning is the major producer of greenhouse gases, energy conservation will have to be a central part of the solution. New York officials can't solve the global warming crisis by themselves. But Abbie Hoffman was probably on to something when he said, "Think globally, act locally."

formaldehyde in the home or office. See table 3.3 for a summary of selected indoor pollutants.

Chances are you first heard of indoor air pollution in connection with the substance radon. This is a radioactive gas, produced naturally from uranium-bearing rocks or soil. Once released underground, radon can migrate into the atmosphere (where it dissipates) or can seep into homes and buildings (where it may accumulate and pose serious health problems). There's no fooling around with this pollutant. Radon has been conclusively linked to lung cancer and is reportedly the second leading cause of the disease, after cigarette smoking. EPA estimates that across the nation, radon is responsible for between 5,000 and 20,000 deaths a year.[38]

How widespread is the radon problem? EPA estimates that nationally perhaps 1 in 10 homes may contain radon at levels warranting remedial action. (In most cases, ventilation systems can be installed in basements or cellars to improve air circulation and bring down indoor radon concentrations.) A New York State Health Department estimate put the number in the state at roughly 1 in 20 homes.[39] Luckily, New York City does not lie within the so-called Reading prong (a radon-containing geological belt that stretches from Pennsylvania, across northwestern New Jersey, into the city's northern suburbs). But government scientists warn that there have not yet been enough radon tests in New York City to make broad generalizations about exposures there. And they caution that even if overall city exposures are quite low, there could still be substantial risks in individual homes.

Are environment objectives working at cross-purposes when it comes to indoor air pollution? Some observers wonder whether energy conservation measures, like tightening building seals, might trap pollutants and raise indoor concentrations. Not to worry. Things like weather stripping, caulking, and sealing lower ventilation rates by about 10 to 30 percent. Such modest reductions produce useful energy savings. However, the changes these measures have on indoor air pollution levels are small, especially when compared to the 10-fold or even 100-fold difference in pollutant concentrations observed between one house and another.[40] Far more effective than fretting about small ventilation losses from improved insulation is to remove offending pollution sources and change the practices (i.e., smoking) responsible for indoor contamination in the first place.

With government agencies having no legal jurisdiction directly over air pollution in private homes and apartments, their major armaments have been public education and restrictions on the manufacture of products or substances that present indoor hazards. They have used the former more than the latter. EPA action to ban domestic fumigation devices containing the pesticide lindane is one of the few instances where the agency has flexed its muscle specifically to help curb indoor air toxins. In another isolated action, the Consumer Product Safety Commission has banned the use of vinyl chloride as an aerosol propellant. And while the Occupational Safety and Health Administration has set some standards (usually not the tightest) for workplace exposures to certain indoor pollutants in industrial settings, it has done little to protect the air office workers breathe.

On the radon front, the federal government's response has been slightly more energetic. EPA has established a federal "action level" for the gas, identifying the indoor radon concentration at which it believes remedial activity is warranted. And under the 1988 Indoor Radon Abatement Act, Congress directed EPA to coordinate federal, state, and local abatement efforts; provide technical assistance; set up regional training centers; and develop model construction standards, among other things.[41]

*   *   *

We end this discussion on an optimistic note. As this book was going to press, Mayor David Dinkins appointed Albert Appleton as the new commissioner of the Department of Environmental Protection. For the first time in years, clear air advocates are finding a sympathetic ear at the agency's helm. The experience of conservationist William K. Reilly, who was tapped by President Bush to head the U.S. EPA, suggests that it often takes more than a change at the top to turn around environmental policies. But around the city's DEP these days, one can't help but detect a fresh breeze blowing.

PART 4

Drinking Water

There is a paradox that threatens the future of New York City's drinking water. Our water system has historically ranked among the nation's best. But this reputation is now handicapping efforts to address brewing problems that all sides concede exist in some form.

It is helpful, in thinking about drinking water in New York, to divide the issue into two camps—water supply (which involves the amount of water available to and consumed by city residents) and water quality. On the supply side, New York City has justifiably basked in the glory of its water delivery system. Three vast reservoir networks, located as far as 100 miles north of Manhattan, funnel roughly 1.5 billion gallons of drinking water into the city and upstate watershed communities every day. Marveling over these expansive holdings, one can begin to see why calls for even further growth of our water supply network are not falling on sympathetic ears outside city limits. And because of our perceived water abundance, the need for comprehensive conservation measures has not, except in times of drought, been wholeheartedly embraced by New Yorkers or their political leaders.

But it is not alarmist to claim that New York City's water supply is

131

in jeopardy. Water consumption from city reservoirs has climbed 30 percent since 1960. Supplies could be thinned even further if upstate communities are allowed to dip deeper into the city's water coffers. Experts also warn that current consumption rates are several hundred million gallons a day more than our reservoirs can safely provide during severe droughts. Such concerns are tempting city water officials to seize new supply sources, despite potentially huge environmental, logistical, and political costs, and before water conservation measures are fully implemented.

The issue of water quality is also plagued with contradictions. For decades, New Yorkers have enjoyed drinking water of superior purity and taste. Our water rarely violates federal or state health standards. And so high has city water quality been that advanced filtering and treatment methods commonly used in water systems around the nation have not yet been employed here.

Rapid development on lands buffering the city's upstate water supply, however, is already undermining its quality. Bacterial contamination from sewage discharges into watershed tributaries (along with road salts, agricultural runoff, and acid rain) is the leading spoiler. And incursions by the city into new supply sources such as the Hudson River could further compromise New York's highly acclaimed drinking water.

But at least until recently, water quality issues have taken a backseat to supply issues in government circles. City officials have not asserted their rightful authority to protect the upstate watersheds from ill-suited development. Some water planners even seem ready to accept declines in drinking water purity as the price for insuring an adequate supply. It is only because our water has historically been of such prizewinning quality that some officials are now willing to consider such a tradeoff.

WATER SUPPLY

Background

New York City's water supply system differs from most. Except in times of drought, it does not rely on a nearby river. Other than for a portion of Queens, it does not depend on underground sources. It is centered instead on three upstate reservoir systems that have been assembled over the past 150 years. The story of the city's water supply is filled with red-letter dates. Here's a peek into history.

1667   Manhattan's first public drinking water well is dug by the British at Fort James on the southern edge of the island.

1750   Some public wells serving Manhattan businesses and residents are becoming foul, a result of cesspool seepage and street runoff.

1835   With two-thirds of city residents still dependent on well waters, New York City voters approve a plan to draw pure waters from the Croton River in Westchester and Putnam counties.

1842   The first upstate drinking water arrives in New York City by aqueduct (from the Croton watershed). Its arrival at the Murray Hill reservoir (at Fifth Avenue and 42nd Street, now the public library) triggers the so-called great water celebration of 1842.

1866   The city adds a second dam to the Croton River, and for the next five decades continues to expand and rebuild the Croton system. During that period, it serves as the city's primary water supply.

1905   With declining water pressure in Manhattan and periodic water shortages in Brooklyn, the state legislature authorizes the city to develop upstate watersheds— providing that the city makes drinking water available to upstate communities.

1927   The city completes construction of the Catskill system (including the vast Ashokan reservoir on Esopus Creek), a 92-mile Catskill aqueduct, and city water tunnel number one (through which Catskill waters begin to flow into New York City).

1931   The U.S. Supreme Court, ruling in a case brought by New Jersey against New York, allows the city to divert up to 440 million gallons a day from the Delaware River, so long as it maintains flows to protect downstream fisheries and downriver water systems in New Jersey and Pennsylvania.

1954   Amid continued interstate squabbling, the U.S. Supreme Court modifies its earlier ruling and raises the city's allocation of Delaware waters to 800 million gallons a day.

1965   Workers complete construction of the Cannonsville reservoir, the final link in the Delaware system and the last reservoir added to the city's supply network.[1]

The System Today

The water supply people like to describe their system in superlatives. *Priceless* and *irreplaceable* are words that roll off their tongues. They aren't exaggerating. New York City's most valuable capital asset could well be its drinking water supply. The system stretches across nine upstate counties. Its three watersheds—the Croton, the Catskill, and the Delaware—cover almost 2,000 square miles, more than six times the size of New York City itself. And its 18 collecting reservoirs and other distribution facilities range from the Rondout reservoir in Ulster and Sullivan Counties (through which nearly one-half of the city's total supply passes) to the Central Park reservoir (which surprisingly to many is still used intermittently for some water destined for portions of Manhattan's Upper East Side and the Bronx). The roughly 550 billion gallons the system holds at full capacity could theoretically satisfy the city's thirst for drinking water for an entire year.

Aqueducts and tunnels are the lesser-known stars of the city's water supply show. Normally, about 90 percent of our drinking water can be traced to the Catskill and Delaware watersheds. After collecting in six huge reservoirs in upstate Delaware, Schoharie, Greene, Sullivan, and Ulster counties, this water travels southeast in two major aqueducts. The aqueducts are grade-level, covered trenches for much of their 90-or-so-mile journey, and underground pressure tunnels for about a quarter of their length. They are constructed mostly of concrete and extend from about 13 to 18 feet in diameter.

How do city waters from upstate watersheds cross the Hudson River? The two aqueducts plunge more than 1,100 feet under the Hudson before veering south in Westchester County. Waters from

Joe Traver, New York Times Pictures

PRICELESS AND IRREPLACEABLE. *New York City's drinking water system is perhaps its single most valuable capital asset. The city's upstate Ashokan reservoir, pictured above, is one of 18 reservoirs that bring roughly 1.5 billion gallons of high-quality drinking water to city taps every day. Despite emerging threats to long-term water purity, the system remains the envy of drinking water providers around the nation.*

the aqueducts mix in the Kensico and Hillview reservoirs and parade into town through city water tunnels number one and two. These tunnels, cut in deep rock 200 to 800 feet below the surface, transport water directly to the boroughs. They are 11 to 17 feet wide, 18 to 20 miles long, and have been gushing with water almost continuously since they were opened more than 50 years ago.

Compared with the extended supply lines of the Delaware and Catskill watersheds, shipping water from the nearby Croton system is a snap. Water from this Westchester County watershed ends up at the New Croton reservoir. It then races 24 miles via the New Croton aqueduct to the Jerome Park distribution reservoir in the Bronx, and from there through the Central Park reservoir in Manhattan.

FIGURE 4.1: NEW YORK CITY'S
DRINKING WATER SUPPLY

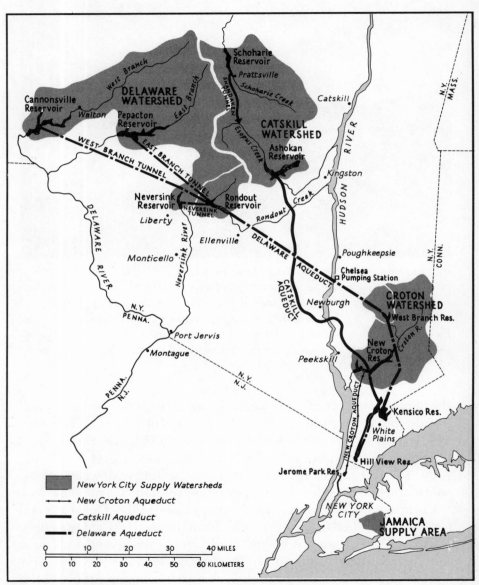

Citizens Union Foundation

FIGURE 4.2: NEW YORK CITY'S
DRINKING WATER DISTRIBUTION NETWORK

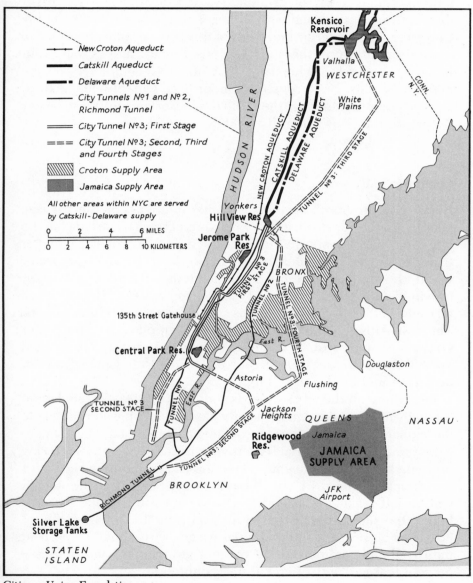

Citizens Union Foundation

The city's far-flung water supply network stretches to three additional outposts. One is the Chelsea pumping station. Sixty-five miles north of the city, it has supplied New Yorkers with up to 100 million gallons a day of drinking water from the Hudson River during three past droughts. Another is the Jamaica Water Supply Company. The underground wells of this private water service still serve residents in southeastern Queens. Finally, there's water tunnel number three; when the first stage is completed in the mid-1990s, this giant construction project will provide a third aquatic highway linking the Hillview reservoir in Westchester County to the city's water distribution network.[2]

Consumption

Every city program seems to have its number. For the garbage people, it's the approximately 19,000 tons of trash disposed of in the city every day. For transit buffs, it's the 5.3 million daily passenger trips on the city's bus and subway network. In the drinking water world, the number is around 1.5 billion tons—the amount of water provided daily by New York City's reservoir system. The city itself takes more than 90 percent of this cut, just over 1.4 billion gallons. The rest—roughly 110 million gallons—supplies drinking water systems in four upstate counties (Orange, Putnam, Ulster, and Westchester). In fact, 90 percent of Westchester County's water needs are met by the city's supply.

Just where all the city's water is going is a matter of speculation. Virtually every major American city has hooked up water meters to individual homes and offices, providing a running tab of daily consumption. Not New York. Its ten-year residential water-metering program has only just begun. (Commercial and industrial accounts are already metered.) Until all home meters are installed, officials will be left making educated guesses about water usage patterns here. One such attempt, based on data from 1983–84, attributes just over 53 percent to residential usage, with about 17 percent industrial and commercial, 14 percent government and public, 9 percent leakage, and 7 percent unaccounted for.[3]

How much water does the average New Yorker use every day?

City officials wish they had a concrete answer to this one, too. About all they can say is that dividing the city's daily water consumption by its population yields a usage figure of just over 200 gallons per person per day. But this number is deceiving. It includes water used in commercial and industrial facilities, and water lost through leakage. A more useful number would be the actual consumption by individual homeowners and apartment dwellers. National figures suggest a per capita usage of about 80 gallons per day just in the home; how much these numbers mean for New York City is anybody's guess.[4]

This much seems evident—water demand here has continued to mount. Records show total water system usage climbing almost continuously since the turn of the century. One reason is rising per capita consumption. From 1960 to 1980, for example, even as the city's official population reportedly dipped by 700,000, total water consumption grew by 350 million gallons a day. (The trend continued at a somewhat slower pace during the 1980s.)[5] This is a somewhat unexplained phenomenon. Experts have different hunches— increased leakage from aging water mains under streets and pipes in older buildings, higher reliance on water-consuming appliances, growing demand from water-cooled office air conditioners, and more visitors and tourists. (Some also question the accuracy of the census count itself.)

Then there are the upstate counties, which have been planting longer straws in the city's water pipeline. Upstate usage has roughly doubled since 1960. Further increases are likely, with a more than 40 percent growth in demand expected over the next 20 years from Dutchess County alone.[6] (There is even occasional talk of funneling water from the city's supply to its downstate neighbors, the overburdened groundwater sources on Long Island.)

The greenhouse effect is another factor that could accelerate these trends. Down the road, planetary warming could raise the Hudson River salt front (jeopardizing upstate drinking water supplies) and cause ocean water intrusion into groundwater (threatening water systems on Long Island). Such events could turn up the heat on the city to allocate more reservoir water to neighboring counties throughout the region.

For some of these reasons, official water watchers have been

**FIGURE 4.3: NEW YORK CITY DRINKING WATER
CONSUMPTION LEVELS (1900–1988)**

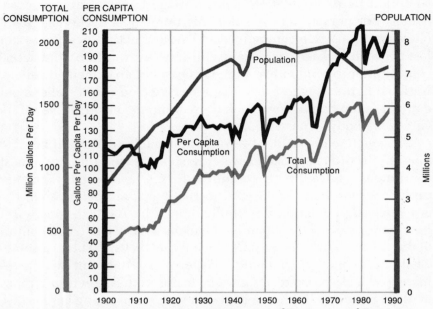

Citizens Union Foundation; New York City Department of Environmental Protection

sounding an alarm. One state planning body, the Water Resources
Planning Council, has projected that even with savings from conservation programs, the system's daily water consumption, under the
worst-case scenario, could rise by over 20 percent, or roughly 1,820
million gallons a day, by the year 2000. In 2030, they speculate, this
number could be as high as 1,970 million gallons a day.[7] Of course,
there is plenty of uncertainty with such long-range projections.

Fear of drought is the major reason why these numbers scare
water planners. During years of normal precipitation, the reservoirs can satisfy existing demand and then some. But every so often
rain and snowfall in the watersheds are below average and the
system cannot be counted on to satisfy regular demand. There have
been three such droughts in the last 25 years—1963–67, 1980–81,
and 1985.

Experts have calculated the amount of water New York's reservoirs can provide during the most severe drought periods. For New York City, this number is just under 1.3 billion gallons a day. If you find yourself at a conference on water planning, you can impress your hosts by referring to this figure as the "safe yield." This means that if a severe drought occurred this year, New Yorkers might have to reduce consumption from the regular yield (of 1.5 billion gallons) by at least 200 million gallons a day.

The Law: Water Supply

There are not many environmental laws that are 85 years old. But a New York State statute enacted at the turn of the century still stands as the dominant law that has shaped New York City's water supply. The 1905 Water Supply Act, and later amendments, effectively gave the city authority to take upstate lands and flood them, creating reservoir systems in the Catskills and Delaware watersheds. In return, the city was required to furnish water to upstate counties in which the reservoirs or water-transporting aqueducts and tunnels were located.[8]

The 1905 law may have served the city well in its battles to secure watershed lands from some upstate interests. But when it came to diverting the headwaters of the Delaware River to New York City from New Jersey and Pennsylvania, it took the U.S. Supreme Court to sort out the legal rights. In a 1954 ruling, the Court modified an earlier agreement between New York and New Jersey, and capped the city's usage of Delaware water at 800 million gallons a day, thereby insuring a steady flow of Delaware water to downstream states.[9]

The law gets a lot simpler when you turn from water supply to water conservation. With federal statutes silent in this area, the issue of how to make the most of existing water supplies has been left to state and local governments. They haven't jumped at the opportunity. The most significant law to date is a 1989 city statute requiring, as of 1992, that all newly built or renovated homes, apartments, and commercial establishments use water-conserving toilets (with 1.6-gallon flushes replacing the 3.5- to 8-gallon norm of standard models). Even though this law does not require the replacement of existing toilets, it is one of the more progressive water conservation measures around. Still, except for emergency regulations applied during previous droughts, there is no comprehensive statute on the books requiring the city to limit overall water usage.[10]

Government Action

It's a little eerie. There is a striking similarity in how government has been responding to New York City's water supply shortfall and to its solid waste crisis. On the garbage front, the city has been seeking to increase disposal capacity with new trash-burning incinerators; it has devoted considerably less energy to recycling and waste reduction. Likewise, the city has been gearing up to tap new sources of drinking water, without first implementing comprehensive water conservation and leak prevention measures.

NEW SUPPLY

Government's battle plan for expanding water supplies is being quietly shaped by a little-known group of local, state, and federal representatives who have been coming together as the Mayor's Intergovernmental Task Force on New York City Water Supply Needs. The city's Department of Environmental Protection, along with the state Health and Environmental Conservation departments, are the main players in this assembly. And reports by the Intergovernmental Task Force have spelled out in one place the city's daring and controversial options for further enlarging New York's drinking water supply.[11]

The Hudson River is government's unofficial first choice. Already the city has pumped 100 million gallons a day into its drinking water system from the Hudson in times of drought. These withdrawals take place at the city's Chelsea pumping station, in the Dutchess County town of Wappingers Falls, 65 miles north of the city. City water officials would like this to become a full-time arrangement.

The advantages include the proximity of the Hudson to New York City's supply lines and the avoidance of major land acquisition costs.

With such benefits in mind, the city has commissioned a study to explore withdrawing Hudson River water in amounts up to 1.2 billion gallons a day, roughly 80 percent of the city's current total usage.[12] But one guesses that city officials might settle for a 200- to 300-million-gallon-a-day Hudson River supplement, itself a diversion of unprecedented scope.

The city's Hudson River water supply of the twenty-first century could take one of two forms. First, it could stick with a traditional pumping station like Chelsea, or expand that plant itself. Such options could yield in the range of 100 to 300 million gallons a day. Alternatively, the city could dust off the so-called high-flow skimming idea, which was first proposed by the U.S. Army Corps of Engineers in the 1970s. Under one approach, it could suck up as much as 1 billion gallons a day or more of Hudson River water near Hyde Park during periods of peak river flows. This water would then be pumped to new or existing reservoirs in that area and transported more than 100 miles to the city's existing water system.[13]

The Hudson River options, still in early planning stages, are already sailing into choppy waters. Upstate activists, like those with Scenic Hudson, point out risks to drinking water supplies in the city of Poughkeepsie and other communities if New York City moves ahead with large-scale Hudson River diversions. (These withdrawals could allow ocean tides to creep northward, subjecting drinking water intake pipes to possible saltwater contamination.) And environmental scientists warn of adverse consequences, among them the disruption of the sensitive, productive, but understudied Hudson River ecosystem. Any permanent Hudson River program could also inject troubling drinking water quality and treatment issues into the debate.

Believe it or not, government officials are seriously considering New York City groundwater as another water supply extender. Beneath portions of Brooklyn and Queens lie layers of aquifers, zones of water up to 800 feet below the surface. At one time, such aquifers supplied most of the water consumed in these boroughs. (But over the years, reliance on the aquifers fell off as a result of overusage and water quality problems.) Today, about 500,000 residents in southeastern Queens are still served by the underground wells of the Jamaica Water Supply Company.

The city's groundwater options, all of which would require the sinking of new wells, are still quite tentative. One alternative envisions the withdrawal of 50 to 150 million gallons a day on a periodic basis. Another would pump out larger quantities (say 200 to 400 million gallons a day) for several months, and then recharge these groundwater sources with water from upstate reservoirs.[14] Both alternatives bring out the yellow flag from those concerned about water quality. (There is even the possibility that New York City and Long Island authorities could reach some agreement on a partial merger of upstate reservoir and underground Long Island supplies. This is a real long shot.)

At least for now, government officials have put aside other long-term dreams for additional supply. The establishment of new upstate reservoirs—one alternative mentioned from time to time—raises giant political and logistical hurdles. Tapping into the Great Lakes and building an aqueduct across the state to transport water of uncertain quality present even greater obstacles. And say good-bye for now to reusing municipal wastewater (too many health risks) and to the desalinization of ocean water (no technology economically available for the scale needed).

CONSERVATION AND LEAK PREVENTION

There's no simple way to report on the city's water conservation program. There are few legal mandates setting conservation targets or binding reductions. And except in times of drought, the city seems to have no clearly stated objective as to what it expects from conservation measures. (We use the term *conservation* here to include both strategies that cut usage by individuals and businesses and those that reduce water losses through systemwide leak prevention and management.)

One benchmark, now being advanced by some environmental advocates, is simply whether the city is squeezing the maximum savings out of its existing water holdings before turning to new sources of supply. Sooner or later, city officials will have to answer this question directly and convincingly.

The launching pad for any water conservation program is metering. Unless you know where water is being used, it is difficult to know where conservation strategies can best be applied. City officials understand this. And it is one reason why in 1988 they finally began installing over 630,000 water meters in homes and apartment buildings, a program they project will take a decade. As of late-1989, roughly 60,000 such meters had been installed. (At the time this program began, there were roughly 180,000 metered accounts—most of them commercial and industrial.) Water officials optimistically estimate that when metering is completed, savings of up to 225 to 300 million gallons a day or more will be possible. This amounts to as much as one-fifth of total city water consumption.[15]

Another thing city officials are thinking about these days is the cost of water. Right now it is still a bargain. In most cases, minimal water charges are assessed on the basis of property frontage, not on actual consumption; households or businesses that are metered are charged only 95 cents per 750 gallons (sewer fees are extra). Metering, however, sets the stage for the city to assess water fees based on true consumption. This could lead to a more equitable billing system, while providing a financial incentive for conservation. Even before universal metering, expect water and sewer rates, which have increased in recent years, to climb further. (By the way, it's the state-created New York City Water Board, controlled by Mayoral representatives, that quietly sets city water and sewer charges.)

If the city can point to modest progress on water metering and perhaps even pricing, it has less to brag about across the rest of the conservation spectrum. One promising strategy—modifying or even replacing conventional toilets, sinks, and faucets so that they use less water—is going nowhere fast. New York City's estimated 4 million toilets alone are flushing as much as 40 percent of the total household consumption. While a new city law directs the installation of low-flow toilets in new construction and major renovation projects undertaken in 1992 and beyond, officials calculate that it will take two decades before this statute yields a projected savings of 200 million gallons a day.[16]

But city officials are not fully exploiting the opportunity to cut consumption from water fixtures in existing houses, apartments, and businesses. This is admittedly a tough assignment. Only a handful of American cities have embarked on such a program. But in San José, California, for example, one city agency distributed low-flow shower heads and water-saving toilet devices to 213,000 households, virtually all of the city's residences. Follow-up surveys showed that more than four out of five homes were utilizing the hardware. These measures alone are netting residential water reductions of 10 percent. And in New York City, two city agencies and the Plumbing Foundation, a trade association, are conducting a pilot program that has come up with some interesting results: a 20-unit building in which low-flow toilets were installed has shaved water consumption by an estimated 30 to 40 percent over similar-sized buildings using only traditional plumbing. Except for such experiments, New York City continues to rely on less-ambitious public education and utility company campaigns to get this job under way.[17]

Other potential, if more radical, programs to save water are rarely being discussed. One such idea is to switch to nonpotable sources (i.e. river or well water) for some uses such as firefighting. Another is to explore the possibility of water recycling systems. (We are told that some Japanese bathrooms use sink water to fill toilet tanks.) It's time for some creative thinking by the city's Department of Environmental Protection on these sorts of concepts.

The most hotly contested issue in the area of water conservation may be leakage. The city's own limited data show a loss of about 10 percent of water every day from leaky city mains. (It also appears as if the city's "leakage" figure does not include leaks from all 8,000 miles of service connections. Those are the pipes that lead from city mains to individual households.) Officials also classify another 7 to 8 percent of city water as "unaccounted for." To be sure, this latter figure includes water used for firefighting and street cleaning. But it presumably also encompasses, for example, water losses from hundreds of thousands of housing units that are dilapidated or have substandard plumbing. Thus, total system leakage is almost certainly more than 10 percent; how much more, nobody really knows.[18]

City officials are particularly defensive here. For one thing, they are knee-deep in the multi-billion-dollar construction of a third city water tunnel. (Among other benefits, this new transporter will allow city inspectors to check for leaks in the two older tunnels.)

Courtesy of Elger Plumbingware

TRENDY TOILETS. *At up to eight gallons a flush, toilets account for roughly 40 percent of household water usage. A new city law, effective in 1992, mandates that low-flow toilets (using no more than 1.6 gallons a flush) must be installed in newly constructed or renovated homes and businesses. These new models, two of which are pictured above, along with water-conserving shower heads and other hardware, could be the best solution to the city's water supply crisis.*

They are also watching over more than 6,000 miles of water mains. Some of the pipes are more than 100 years old, and detecting underground leaks is not easy work. Water officials talk proudly about their state-of-the-art, electronic leak detection units, whose operators cruise around town scanning for water main cracks.

But the city's leak work leaves room for improvement. Public complaints about water leaks, the number of actual main breaks, and the response time to street leaks all rose from fiscal year 1988 to 1989. And the Mayor's Intergovernmental Task Force notes that individual city water lines are surveyed for leaks only once every six years. Speeding up this maintenance cycle, it suggests, could cut the leakage rate by nearly 50 percent.[19] No surprise then that the issue of water leaks continues to hound department officials.

Ironically, city efforts have already shown the promise of conservation. For example, during one month in the 1985 drought, water consumption dropped by over 300 million gallons a day.[20] The conservation measures that triggered this short-term fall-off may be different from those needed for year-round savings. (The city's emergency drought regulations include everything from restrictions on car washing to reduced hours for office air conditioning.) But the success of city drought plans suggests that New York's seemingly insatiable water consumption may be more elastic than it first appears.

WATER QUALITY

Background

The issue of water quality in New York City may be different from what some people think. For those who worry about such things, the occasional chlorine odor, muddiness, or off taste of their tapwater, is what is most troubling. But government officials see things differently. Their biggest concern is delivering water free of bacteria and other disease-causing organisms. This has historically been the primary target of drinking water providers. While typhoid and cholera threatened city supplies in the 1800s, these days it is things like the parasite *Giardia lamblia* (a microorganism that can cause severe intestinal problems) that has planners on guard.

Toxins in our water supply are a secondary target of public health experts. Fortunately, New York City supplies have largely escaped the toxic fouling that has made headlines on Long Island and elsewhere. There are two reasons for this—our primary reliance on upstate reservoirs (not groundwater) and the rural nature of our watersheds (although this is becoming less true). So while New Yorkers may be most concerned about what they see or taste in their tapwater, it is the hidden bacterial and toxic agents against whom most water quality battles will be waged in the 1990s and beyond.

The Law: Water Quality

You could sort the laws covering New York City water quality into two stacks. If you did, one would include statutes designed to protect the sources of our drinking water. The second would contain laws to limit contaminants in the water that ultimately flows into New York City.

The federal Clean Water Act, perhaps best known for its edict to make the nation's rivers fishable and swimmable, also has a say when it comes to protecting the city's drinking water supply. The statute regulates the discharge of pollutants into all navigable bodies of water, including New York's reservoirs and their tributaries. Municipalities, private institutions, and anyone else seeking to jettison wastes into these sources must obtain state permits that limit the quantities and concentrations of pollutants they may discharge.[21]

Even more effective than limiting pollution discharges into drinking water sources is controlling watershed development in the first place. That's why a state law granting the city broad powers to regulate watershed uses is potentially so important. Under the 1909 statute, since amended, New York City drew up regulations that place restrictions on some public uses and activities in the reservoirs and on their surrounding tracts. But these rules, last modified by the city in 1953, are hopelessly outdated.[22]

The next group of statutes drops a second line of defense. Under federal law, all public water systems must, in addition to other duties, insure that the water they deliver does not exceed maximum contaminant levels for certain bacteria and toxins, and meets other standards governing such things as acidity and odor. This 1974 Safe Drinking Water Act mandate prompted the U.S. Environmental Protection Agency to set contaminant levels for 30 pollutants. Amendments passed in 1986 are compelling the agency to more than triple the number of substances so regulated.[23]

Another provision in the 1986 federal statute, designed to clean up drinking water before it reaches the city, has aroused our municipal officials more than any other law on this subject. It directs EPA to define the circumstances under which surface drinking water systems like New York City's must install expensive water-filtering equipment. Under EPA's 1989 rules implementing the directive, municipalities may obtain an exemption from the filtration requirement only if they can demonstrate that they have put in place a stringent watershed control program and meet tough water quality standards.[24] (The New York State Health Department has the important task of enforcing all these federal laws, along with its own regulations adopted under separate state statutes.)

Water Quality Today

New York City's drinking water on the whole remains unusually high in quality, at least for the present time. It consistently meets virtually all state and federal health standards, surpassing many criteria by wide margins. Some of the best news may be the absence of any chronic heavy metal, or other toxic contamination, although some observers are beginning to worry about pesticide runoff in the watershed. On taste alone, New York City tapwater has been ranked as one of the best in the country, equaling or surpassing that of many fancy bottled waters.[25] And one guesses that city managers around the nation would be delighted to swap their own water supply systems for New York City's reservoirs.

Which neighborhoods receive the best water in New York City? This is one question local water quality experts are frequently asked. Normally, about 90 percent of the water delivered to New Yorkers comes from the city's two highest-quality sources—the upstate Catskill and Delaware watersheds. Water from these two systems, which is mixed together before reaching the city, is ultimately delivered to most of Manhattan, Queens, and the Bronx, and to the entire boroughs of Brooklyn and Staten Island.

Water from the nearby Croton watershed in Westchester County, which is several notches below in quality, is distributed to portions of northern Manhattan, the Lower East Side, a narrow strip along the Hudson, and parts of the Upper East Side, as well as sections of the south and east Bronx. (This allocation is largely a result of geography. The neighborhoods now receiving Croton water were part of the original service area of the old Croton system and are located in low-lying portions of the city, which are able to receive lower-pressure Croton water without the need for costly pumping.)

For now, the differences in quality in water drawn from the three reservoir systems are still subtle. There is presently no cause for panic even for consumers drinking water exclusively from the Croton system. The Croton's deterioration is showing itself primarily through hazy signs, including increased turbidity, or cloudiness, and occasional algae blooms from sewage and other organic wastes. The real danger is not consumption of Croton water today, but the possibility that unchecked declines will lead to even more serious problems down the road.

Residents of southeastern Queens also have justification for concern. Their drinking water comes primarily from underground water tables (the Brooklyn-Queens aquifer), which are not up to the purity of the upstate reservoir systems (see page 152).

New York City has maintained its high-quality water with a little help from its friends. Significantly, city water keepers allow water to remain in the reservoirs for up to 12 months under normal conditions, before sending it into the city; this long residence time acts to purify the water, allowing for the natural fallout of solid matter. City officials also keep an eye on how water quality in the different reservoirs compares. And by making selective withdrawal, they seek to rely most on water of the highest quality.

Then there are the additives. The city's water troops pour chlorine into the distribution network to kill bacteria and other disease-carrying organisms. They occasionally add alum (an aluminum compound) and copper sulfate (a water-soluble salt) to control turbidity and algae growth, respectively, in all three reservoir systems. Same for caustic soda, a chemical that reduces acidity. (And they inject fluoride, not to control quality but as an aid in dental care.)

Both chlorine and alum have their drawbacks. When chlorine is used to disinfect water, it reacts with organic matter to form trihalomethanes, a group of compounds that are suspected carcinogens. These troubling compounds are found regularly in trace amounts in city water, although below state and federal drinking water limits. As to alum, fishery groups warn that one long-term impact of adding this chemical to the reservoirs is an accumulation of sediments along reservoir floors, smothering small aquatic organisms that live there.[26] City water experts respond that, on balance, the benefits of these chemical treatments outweigh the risks.

WORRISOME WELLS IN JAMAICA, QUEENS

The city's illustrious water supply family has an ill-treated cousin—the private water company serving the residents of Jamaica, Queens. Unlike the rest of the city's system, the Jamaica Water Supply Company draws upon underground water tables, not upstate reservoirs. Unlike the rest of the city's system, some of the Jamaica Water Supply sources have been closed due to toxic contamination. And unlike the rest of the city's system, Jamaica Water Supply drinking water requires greater chemical treatment and is of lesser (although apparently still acceptable) quality than the water consumed by New Yorkers in the rest of the city.

The Jamaica Water Supply Company is unique in at least one other respect—it is the last private water provider in New York City. It supplies, on average, about 30 million gallons a day of well waters. Its Queens service area extends west from the Nassau County line, south of Grand Central Parkway to Kennedy Air-port, and east of a line one mile west of the Van Wyck Expressway. Approximately 500,000 residents of southeastern Queens (along with 120,000 in adjoining Nassau County) are its customers. (They don't have much choice; there's no competition in the water supply business.)

The big problem with Jamaica water is quality. The company owns 75 wells that bore down through soil and clay into four water-storing layers of the Brooklyn-Queens aquifer. But as of mid-1989, 20 of these wells were closed by order of the New York City Department of Health, due to the presence of toxic contaminants. Among the organic chemicals that have surfaced are trihalomethanes (which are formed from the addition of chlorine as a disinfectant) and tetra- and tri-chloroethylene (synthetic solvents and degreasers, which were probably stored or disposed of improperly above the aquifer and which later percolated into the earth).[28]

The natural chemistry of the Brooklyn-Queens aquifer doesn't help much. Its water-bearing soils, like those of neighboring Long Island, are somewhat acidic, with high levels of iron and manganese. In part for these reasons, water from the Jamaica wells may at times be off taste or off color.[29]

What Queens residents want most to know is whether Jamaica water is safe to drink. According to city officials, the answer is yes. The city Health Department samples the water from the system's wellheads, distribution points, and taps six times a year and conducts extensive tests. (The monitoring began in 1979 after neighboring Nassau County had detected organic chemicals in some wells whose underground supplies are basically contiguous with the Jamaica system.) If a problem is identified, the well in question may be closed.

Public health officials acknowledge that Jamaica water supply waters are of lesser quality than those from upstate reservoirs, and that they often contain organics, at least in trace quantities. But they maintain that all open wells meet state water quality guidelines and that their current monitoring program keeps them on top of the situation. (The city has been supplementing the Jamaica company's supplies with upstate reservoir water to the tune of 30 million gallons a day. These reservoir waters are not mixed with the system's well supplies in any equal apportionment; neighborhoods in Jamaica receive either all reservoir water, all well water, or a mixture of the two, depending on their location in relation to the city's regular water distribution network.)

The presence of organic chemicals and (of all things) water rates that were higher for Jamaica residents than for New Yorkers supplied with city water prompted state legislative action. A 1986 statute directed the city to assume control over the privately owned water utility, by condemnation if necessary.[30] In mid-1989, negotiations and legal maneuvering for a city take-over were continuing. So was the essential water quality monitoring.

Eric A. Goldstein

HOMEGROWN WATER. *Unlike other New Yorkers, about 500,000 residents in southeastern Queens are drinking water drawn from underground aquifers. The Jamaica Water Supply Company is the last private water provider in New York City. But water from its wells, stored in tanks like the one above, is of lesser quality than water from the city's reservoirs.*

The weakest link in the water delivery system is the end-of-the-line plumbing. Once water leaves city-owned mains, it travels through private service connector pipes, rooftop water tanks, and interior plumbing. It is on this final journey that water can pick up such things as lead, rust, and taste problems. In fact, more than one in ten New York City households may have tapwater lead levels that exceed federal health standards.[27] City officials, their hands already full, seem less sympathetic to such troubles.

Threats to Water Quality

Very little of the city's tapwater comes from rain falling directly into our upstate reservoirs. (Officials guess perhaps no more than 3 percent.) Water supplies are primarily replenished as rain and melting snow run off mountains and flow into tributaries that in turn spill into the reservoirs. For this reason, environmental conditions in the 2,000 square miles of watershed lands surrounding the reservoirs should be of paramount concern to all who care about the quality of city drinking water.

Today, New York City's watersheds are under attack. The main opponent is inappropriate development. Until the last few decades, the three upstate watersheds were, for the most part, rural and sparsely developed. No longer. The nature of the city's watersheds, especially the Croton, is irrevocably changing as suburban and second homes and even shopping malls and office buildings fan out into upstate counties through the region. The population of West-chester, Putnam, Ulster, and Sullivan counties—home to most of the city's upstate water reserves—increased by 20 percent from 1960 to 1980. The highly regarded Regional Plan Association estimates that this number will jump another 15 percent by 2015.[31] With a growing populace comes the problem of sewage.

It might seem crazy to allow sewage systems to discharge their wastes into city drinking water sources. But New York State continues to sanction such practices, which are legal under the Clean Water Act. As of mid-1989, 83 sewage treatment plants were discharging into tributaries that feed New York's reservoirs. Twenty-nine more sewage facilities are being planned, reports environmental advocate Robert F. Kennedy, Jr., and his students at Pace University Law School, who are becoming the unofficial protectors of the city's watershed. Total discharges here were calculated at more than 10 million gallons a day. (Connecticut and Rhode Island have both prohibited such discharges under state law.)[32]

The Croton system lies most directly in the path of advancing development, and it has been hardest hit. More than two-thirds of the permitted pollution dischargers use Croton tributaries as their receiving waters. To make matters worse, a number of these polluters regularly violate their discharge permits. According to one recent city review, among the handful of worst violators are Putnam Hospital (which discharges potentially disease-causing wastes directly into the city's Croton Falls reservoir) and a state prison, the Bedford Hills Correctional Facility (whose effluent, which has regularly exceeded permit limits for sewage-related wastes, flows into a tributary just upstream of the city's Muscoot reservoir).[33]

Ironically, the city itself owns upstate sewage plants that discharge into the watershed. In fact, its seven facilities contribute 30 percent of the wastewater flow from all watershed treatment works. (The city originally built these plants under state law, to insure that sewage waste from small upstate communities would not contaminate the city's water supply.) While the city concedes that all these plants need reconstruction, the 50-year-old Mahopac sewage works in Putnam County may be in the most desperate shape. Records of its discharges showed a sky-high 86 percent noncompliance rate with its permit limitations in 1988. The impacts of such discharges are already becoming evident to those who survey water quality in the Croton. No wonder the city now plans to install filtration equipment (at a cost of roughly $600 million) to help purify the Croton's supply.[34]

It's hard enough monitoring water pollution sources that have their own discharge pipes. Things get even tougher when it comes to runoff, the catch-all term that covers all kinds of contaminants that wash off roads, driveways, and embankments, into reservoirs and their tributaries. Road salts used for winter deicing, gasoline, oils, and fertilizers are some of the common intruders. (Technical buffs like to call this "nonpoint source" pollution.) Reservoir keepers in the Delaware system have encountered an unusual

Richard Knabel

SEWAGE IN THE WATERSHED. *Every day, more than 10 million gallons of sewage wastewaters are dumped into tributaries that feed the city's upstate reservoirs. The Mahopac sewage treatment works, pictured above, is one of 83 such facilities discharging wastes in the upstate watershed. To make matters worse, this plant and others regularly violate pollution limits in their state permits.*

aspect of this problem—runoff from dairy farms. Fertilizers and manure have led to an overload of nutrients flowing into the Cannonsville reservoir, prompting water officials to halt withdrawals from that source for several months during past years.[35]

Some threats to the quality of the city's drinking water come from, of all places, the sky. Right now, the most troubling fallout may be acid rain. When coal-burning power plants (many in Midwestern states) discharge pollutants from their tall stacks, the contaminants can travel hundreds of miles and fall to the earth as acidic rain or snow. The Environmental Defense Fund reports that such emissions have contributed, among other things, to levels of sulfates in some upstate reservoirs that have made these waters highly corrosive. Sulfate levels in the watershed are reportedly declining (following controls on sulfur sources at the regional level). But the corrosive powers of existing airborne pollutants is probably increasing the amount of lead, aluminum, and other metals that make their way into our drinking water from pipes and plumbing.[36]

Finally, if the city turns to new sources of drinking water, it might well be opening a can of worms on the issue of quality. Of course, as city officials are quick to note, the Hudson River is already used by upstate communities like Poughkeepsie as their sole source of drinking water. But no one disputes that the Hudson has had troubling PCB contamination and is of lower quality than the city's reservoirs. As for recharging and future withdrawals from the underground Brooklyn-Queens aquifer, keep in mind that some Jamaica Water Supply Company wells, which also draw from the aquifer, are presently closed due to high levels of organic pollutants. The real threat from reliance on the Hudson River or Brooklyn-Queens aquifer is that both sources are more vulnerable to pollution beyond the city's direct control. Once such contamination occurs, it is exceedingly hard to clean up.

Scientists can debate how much, if at all, such new sources would jeopardize public health in the short run. But this aside, the city's decision to tap the Hudson or the aquifer might well be seen as an indicator that New York officials, in an effort to assure new supplies, are willing, apparently for the first time, to accept some decline in water quality.

Government Action: Water Quality

A number of respected water quality professionals have leveled a disconcerting charge against city and state water agencies. They assert that government actions are, in effect, allowing a subtle but irreversible downturn in the quality of city drinking water. Their concern is that unchecked development in the upstate watersheds is lowering water quality. And they suggest that state and city officials have not yet decided whether to demand that watershed development be appropriately planned or to rely instead on after-the-fact filtration (a move they argue is both enormously expensive and potentially less protective of drinking water purity).

Keith Meyers, New York Times Pictures

SUPPLY UP, QUALITY DOWN. *In three previous drought emergencies, the city has activated its Chelsea pumping station to draw water from the Hudson River, 65 miles north of the city. Now, city officials hope to make the Hudson a permanent contributor to the city's water supply. Water chiefs are exploring the possibility of diverting hundreds of millions of gallons a day or more from the Hudson, apparently willing to accept declines in water purity to satisfy growing demand.*

New York City officials take strong exception to these allegations, portions of which appeared in a detailed 1988 exposé in *New York Newsday*.[37] The city's water chiefs argue that their watershed protection powers are limited, that they oppose costly filtration for other than the Croton system (which they are being forced to consider, they say, only in response to federal and state pressure), and that they hope to postpone water filtration for as long as possible. But the critics point first to the watershed protection record of city and state agencies, which most observers would have to admit is thin on accomplishment. A case in point is the city's watershed regulations, last revised more than 30 years ago. Everyone concedes these rules contain inadequate safeguards to protect buffer lands from encroachment. But the city has yet to revamp them.

BOTTLED WATER EVERYWHERE,
BUT IS IT WORTH THE PRICE?

These are heady times in the bottled water business. Nationwide sales jumped 300 percent between 1976 and 1986, according to the International Bottled Water Institute, the industry's trade association. The surge is expected to continue right through the end of this century. New York, which often finds itself at the vanguard of such gastronomical trends, ranks fifth among the 50 states in per capita consumption of H_2O in bottled (and plastic containerized) form.

Two groups of New Yorkers are scooping bottled waters off the shelves of supermarkets around town. One set is motivated by health concerns and sees bottled products as a safer substitute for city tapwater. This group really got charged up after a 1985 plutonium scare (in which trace levels of the radioactive element were detected for a short period of time in city drinking water samples) and the water drought that same year (which prompted city officials to supplement reservoir supplies with withdrawals from the Hudson River). The second group has turned to bottled waters, especially carbonated or sparkling waters, as a no-calorie alternative to soda or alcoholic drinks.

Of course, the interests of the two groups probably overlap. And both may also be turning to bottled water because they believe it tastes better. (But independent water tasters at Consumers Union rated New York City's drinking water supply as "excellent" in separate taste comparisons during 1980 and 1987; they note that city tapwater tastes best when chilled.)[40] New York City's bottled water purchasers don't make their decision primarily on the basis of price; the cost of a 23-ounce bottle of Perrier is more than 4,000 times greater than a similarly sized decanter of New York City drinking water.

Is bottled water worth its cost? As a soda or alcoholic substitute, perhaps. But bottled water may be no better than tapwater in terms of public health and safety, government studies reveal. A 1988 survey by the Suffolk County Department of Health, for example, found that while most bottled waters met current drinking water standards, synthetic organic chemicals (primarily trihalomethanes, trichloroethane, and toluene) were found in almost 50 percent of the brands tested. And according to one source, up to a third of all bottled waters sold around the country are actually taken from local drinking water supplies, most of which are presumably of lower quality than New York City's reservoir-fed system.[41]

Don't count on stronger laws to assure bottled water quality. The statutory framework here, as is the case for public drinking water systems, could hardly be called comprehensive. Insuring the purity of bottled water involves the protection of its source, control of the production line, and enforcement of what government rules and trade association guidelines do exist. But in the absence of legislation or rules that mandate frequent testing and quality controls for bottling and storage, oversight boils down to industry self-monitoring.

Water filters are probably not the answer for New York City residents, either. Consumers Union reports that carbon filters (either attached to the faucet or located underneath the sink) can improve taste and odor and remove organic chemicals, but are not effective against lead and other heavy metals. (The more expensive reverse-osmosis filter systems are better for that task, as are filter cartridges specifically designed to remove lead.) Since New York City tapwater generally is safe from organic chemicals and is of high quality in terms of taste and odor, water filtering devices in most city households are simply unnecessary.

Does buying bottled water ever make sense? Yes. In water emergencies, if high levels of lead have been found in your tapwater, or if you have quality or taste problems that water officials cannot explain, you might want to think about the bottled water alternative. Before doing so, you should consider writing to the bottler for information about the water's source, the steps that are being taken to protect the supply, and the treatment processes, if any. The best advice, if you have any reason to suspect city tapwater, is to notify the Department of Environmental Protection and if its staff are unable to resolve your concerns, to get your water tested independently.

Another problem is enforcement. There are apparently only 25 city inspectors assigned to patrol the 2,000-square-mile watershed. They may be playing a deterrent role. But Department of Environmental Protection staff could not recall the department having pursued a single watershed violation to trial in recent years. (Meanwhile, the city's Department of Health has backed away from its historic position as water watchdogs; the department blames fiscal constraints for having effectively ended its comprehensive watershed surveys.)

The state's record is no better. For example, the Department of Environmental Conservation allows sewage plants to discharge into reservoir tributaries so long as they obtain a state permit. As if that isn't bad enough, the department handles these permits on essentially a case-by-case basis and has yet to review the cumulative impacts of more than 80 such polluters on reservoir water quality. The state Health Department also has plenty of authority when it comes to city drinking water quality. But its primary objective appears to be assuring ample water supplies to communities throughout the state; it has not issued specific rules to protect the city's unique watershed.

Stung by recent attacks on the adequacy of its own drinking water protection efforts, the city unveiled a new water quality preservation program in the summer of 1989. The program's first-year goals include expanded water quality testing and closer monitoring of proposed development projects in the watershed. Long-term plans are aimed at reducing agricultural runoff into the Cannonsville reservoir, upgrading the quality of tributary waters leading to the reservoirs, and acquiring land or protective easements, with state assistance, to buffer the watershed. The city, in announcing this initiative, called it a revised ten-year plan. The decade-long timetable, one suspects, reflects both the complexity of the task and the desire to lower expectations that a lot will change quickly.[38]

Any way you look at it, perhaps the major water quality battle in the 1990s will be fought over filtration. Under new EPA rules, the city and state have a stark choice—either to construct giant filtration plants at the upstate reservoirs or to arrest the degradation in the watershed. While the decision has already been made to go ahead with filtration of the Croton system (a necessary move in light of the failure to control development pressures on the Croton watershed), there has been no clear pronouncement on what to do with the remaining 90 percent of the city's supply. (Among the disadvantages to filtration are: (1) it is enormously expensive, with an estimated price tag of about $3 billion for treatment of Catskill and Delaware waters; (2) the cost of filtration will empty government coffers of funds that could otherwise be used to purchase watershed lands or for other source protection measures; and (3) it could take the pressure off efforts to keep the reservoirs free of contaminants in the first place.)

The good news is that the Catskill and Delaware systems are still of high-enough quality that filtration is not needed now and in the opinion of many may never be required if the city and state take adequate steps to control development and protect the watershed. While the state Health Department has the official say-so in the filtration debate, it is city policies that will determine the final outcome.

Perhaps the boldest response to the quality issue would be city and state purchases of watershed properties. (Presently, about 7 percent of watershed lands are city owned; no figures could be found regarding state ownership.) By buying up endangered parcels that buffer the reservoirs, or their tributaries, officials could help assure the purity of watershed runoff from those properties. Unfortunately, there's no progress to report on this front yet. Governor Mario Cuomo has proposed a $1.9 billion 21st Century Environmental Quality Bond Act that, if passed, would provide funds for securing critical watershed parcels. The fate of this desperately-needed proposal is uncertain. No state environmental bond act funds have been used to secure critical watershed parcels. Nor have city officials, guided by competing economic pulls (as well as political ones), chosen to invest in such purchases.

So far, New Yorkers have been lucky. But as the New York Academy of Medicine, hardly a bunch of wide-eyed radicals, has recently noted, "The fact that there have been no major outbreaks of waterborne diseases in the recent past should not lead to a sense of complacency." The Academy warns, "The immediate need to assure adequate water quantity must not detract from efforts to preserve water quality."[39] The new battle lines are now being drawn.

PART 5

Toxics

I t's not easy to make sense of the toxics problem in New York City. Public reaction to reports of toxic threats to health and the environment often takes one of two forms—panic (buying bottled water in response to concerns over tapwater quality) or indifference ("Everything causes cancer these days, so why worry?"). Such reactions are largely shaped by television reporting, which is where most New Yorkers get much of their news. And the usual government response—that a particular incident presents "no immediate threat"—only adds to the public confusion.

There are subtleties to the toxics problem that are frequently lost in the media translation. For one thing, although a single high-dose exposure to certain toxic substances may lead to permanent health problems, the risk from toxins at lower levels more typical of urban exposures is usually long-term and cumulative. Second, risks spiral from combined exposures to a variety of toxic substances. Such information, usually not included in coverage of the breaking story, does not make a particular toxic occurrence less serious. But it helps place the event in context. The challenge then to communicating information on toxic threats in New York City is to convey the seriousness of the potential danger without causing undue alarm.

Cataloguing every possible toxic threat in New York City and ranking their respective hazards is a daunting task and well beyond the scope of this book. Instead, in this discussion, we survey some of the most significant toxic problems facing the city and attempt to give some form to this amorphous territory.

The number-one toxics problem in New York City today? It is most probably exposure to harmful chemicals in the workplace. The populations at risk range from bridge and tunnel workers to construction trades workers, to dry cleaners to hospital and laboratory workers. For the general public, the major threats are from lead (a toxic heavy metal that poses especially serious problems for children, usually poor, living in homes with leaded paint) and asbestos (a carcinogen present in roughly two-thirds of city buildings and whose deterioration or improper removal can send invisible fibers flying).

Polychlorinated biphenyls (PCBs) might not be the loudest toxic problem screaming for nationwide attention, but they remain a troubling threat in New York, primarily because of their long-standing contamination of the Hudson River. In contrast, hazardous wastes are a quieter topic here than elsewhere, although not nearly as silent as many government officials would have you think.

As policymakers continue to debate the precise risks of dozens of contaminants, the major toxic skirmishes for New York City in the 1990s will likely be fought within the five spheres discussed below.

LEAD

Background

There's a disturbing aspect to the pollution story that goes beyond public health risks and environmental harm. Too often the burdens of pollution fall disproportionately on the poor. Around the globe, you see this in the export of hazardous wastes from industrialized countries to less developed nations. Across the United States, the trend is apparent in the siting of many toxic dumps in low-income communities. In New York, a classic example is the handicap that some city children face from exposure to the toxic metal lead.

That lead pollution persists as an environmental problem just a decade short of the year 2000 is ironic. Unlike most contaminants, lead's dangers have been known for centuries. Pliny, the writer and statesman of ancient Rome, warned his countrymen of the risks from breathing lead vapors. Lead was used in those days as an ingredient in face powders, rouges, and mascaras; a pigment in many paints; a wine preservative; an ingredient in cups, plates, pitchers, and utensils; and as the pipe metal in Rome's vast network of plumbing, which supplied drinking water to the city. Experts now believe that contamination from the widespread use of this malleable metal contributed to unprecedented outbreaks of gout, to sterility and other reproductive problems, and to insidious mental disabilities among the Roman empire's ruling class.[1]

So it is not hard to understand the frustration of physicians in clinics around New York City who continue to see patients suffering from the effects of exposure to lead during the 1990s. Of course, the news here is not all bad. Levels of lead in the city's air, previously a major exposure route, have declined dramatically over the last 15 years—a result of the U.S. Environmental Protection Agency's national phasedown of lead levels in gasoline. But still troubling is high lead exposure from peeling paint in portions of the city's older housing stock. And, along with accumulated lead in dust and soil, lead from old drinking water pipes in some homes, apartments, and other buildings attacks thousands of city residents.

Eradicating the city's lingering lead problem won't be easy. The main offensive—against exposure from old, leaded paint and from drinking water pipes—will require intensive house-to-house combat. But lead-related health problems are fully preventable, and a concerted attack could yield big gains. Government officials must now decide whether the city's poor, who are most susceptible to lead's adverse health effects, will receive the full complement of antipollution resources that city, state, and federal agencies can muster.

Lead: Health Effects

More than 1,900 years after its harmful properties were recognized, lead continues its assault on public health. Its primary victims today are children. Preschool youngsters are highly susceptible to lead's adverse impacts because developing nervous systems are especially vulnerable to toxic agents. Also, children may absorb and retain more ingested lead than adults. The human fetus is also at risk from maternal exposure to lead at toxic levels.

Research studies have demonstrated the dangers from lead at all levels of exposure. In high concentrations, this toxic metal can cause anemia, convulsions, damage to the central nervous system, and, in extreme cases, even death. More commonplace exposures to lead have been associated with learning difficulties, behavioral problems, hyperactivity, and reduced IQ levels in children. Recently, researchers have suggested that significant deficits in mental development can result from prenatal exposure to very low lead levels or from low-level exposure in early childhood. Government scientists have also tied elevated blood lead levels to increased blood pressure in adult males.[2]

Lead in New York City: Exposure Routes, the Law, Government Action

"Lead is potentially toxic wherever it is found, and it is found everywhere," the U.S. Public Health Service recently reported.[3] To one extent or another, New Yorkers must confront lead in paint, dust, soil, drinking water, food, and air.

Those who think the New York lead pollution problem is a thing of the past would do well to check in with the city's Health Department. Some data suggest that lead poisoning has declined in New York City since the early 1970s. Still, random city screening programs found roughly 1,200 children with elevated blood lead levels in fiscal year 1987, nearly 900 children in fiscal year 1988, and projections (based on increased blood sampling) were for 1,300 children in fiscal year 1989. The department's testing net is not thrown over the city's entire preschool population, but extends only to some children in certain high-risk neighborhoods. And the respected federal Centers for Disease Control are likely to lower the level of blood lead believed to represent undue exposure. When such action occurs, the number of New York City children identified as having excessive lead in their systems could jump by ten times or more.[4]

Wide World Photos

TOXIC LEGACY. *As levels of lead in city air have declined over the last two decades, lead in paint remains the most troubling source of exposure for New Yorkers. In 1989, more than 1,300 children, most of whom reside in substandard housing that was painted years ago with lead-containing pigments, were expected to suffer from this highly preventable public health menace.*

Lead in Paint

Leaded paint is today the major source of high-level lead exposure for New York City children. It's not new paint that's the problem. Lead is now virtually prohibited as an ingredient in new interior and exterior household paints sold in the United States. But New York City's large pre–World War II housing stock has left a legacy of peeling and flaking paint, much of which contains high lead levels. Preschool children are the primary group at risk because of their tendency to mouth and chew nonfood items, such as paint chips and objects covered with lead paint or dust. And a single paint chip, less than an inch in size, may contain more than ten times the maximum safe daily level of lead, set by the U.S. Food and Drug Administration.[5]

As if they didn't already have enough to worry about, the city's poor bear the brunt of leaded paint problems in New York. Four low-income areas in Brooklyn—Bedford-Stuyvesant, Brownsville, Bushwick, and Fort Greene—have recorded almost 50 percent of the reported cases of lead poisoning in New York City from 1970 to 1988.[6] This figure reflects the concentration of older and often deteriorated housing in what is sometimes referred to as the city's "lead belt."

Government efforts to limit exposure to lead-based paint began in the early 1970s. In three separate enactments, Congress banned the application of leaded paints to cooking, eating, and drinking utensils, as well as to toys and furniture articles. The single most significant federal step came in 1977. That year, the Consumer Product Safety Commission prohibited, with limited exceptions, the sale of all interior and exterior household paints and coating materials containing more than 0.06 percent lead.[7] Significantly, however, no federal law or regulation mandated the removal or treatment of lead-based paints in existing homes or apartments.

New York City law now provides a further layer of protection, at least in theory. Under a 1982 city mandate, property owners are required to strip or cover lead paint in all dwelling units where children under six years old are living. Another provision of law calls upon city officials to inspect residences in which children who have been identified as having elevated blood lead levels reside.

FIGURE 5.1: NEW YORK CITY'S LEAD BELT

New York City Department of Health: Bureau of Lead Poisoning Control

Where leaded paint is subsequently discovered, landlords must remove or cover it without delay.[8]

Federal actions to reduce exposure to leaded paint have stalled in recent years. On the plus side, restrictions on the lead content of new paint have diminished the threat of lead poisoning for many children. But the federal government has skirted one of the biggest issues—safeguarding youngsters living in the estimated 27 million older housing units nationwide that contain lead-based paints. Also, government regulators have been slow to enforce rules requiring lead paint abatement actions in federally assisted housing. And federal funding for childhood screening programs to detect lead exposure in New York and elsewhere effectively dried up during the 1980s.[9]

With city officials left at the helm, the lead battle has focused primarily on childhood testing rather than paint removal. The number of children screened for elevated blood lead in both public and private programs has nearly tripled since 1970 and was anticipated to reach more than 300,000 in fiscal year 1989. City officials have, however, been less aggressive in enforcing the local law aimed at removing leaded paint from homes in which young children live. According to evidence in one recent legal action, it has taken up to a year to remove lead paint hazards in apartments where children have actually been diagnosed as having lead poisoning.[10] Undoubtedly, many such violations simply go unidentified.

Lead in Drinking Water

Many New Yorkers would be surprised to learn that the water they drink may be a source of toxic lead. This metal rarely occurs naturally in drinking water supplies. But it is sometimes present in the piping and soldered joints of homes and apartment buildings, as well as in plumbing connections that carry drinking water from city mains to individual residences. Corrosion of these pipes and joints can flush potentially harmful levels of lead into New York City tapwater. Lead levels are highest first thing in the morning, after the city's slightly acidic water has been left standing in household pipes overnight.

The extent of lead infiltration into city water is difficult to gauge. Monitoring by the Department of Environmental Protection has found that lead levels at pumping stations and distribution points consistently meet existing federal and state drinking water standards. But these samples are not drawn from household faucets. And unofficial tests by others have uncovered high lead levels in some city tapwater. A random Health Department sampling of so-called first-draw water revealed that 12 to 13 percent of tapwater tested had lead at levels that exceed the existing federal limit. And a small 1988 survey by the Plumbing Foundation of the City of New York, a private trade organization, reached similar conclusions.[11]

In the campaign against lead in drinking water, the U.S. Environmental Protection Agency has been deploying its forces cautiously. Some say too cautiously. In 1986, EPA adopted rules prohibiting the use of lead in solder used for the construction or repair of public drinking water systems and connections. At the same time, the agency also restricted the lead content of new pipes and fittings used for such purposes. EPA believes that its existing lead-in-water standard is too lenient and proposed a tighter limit in 1985.[12] But the agency has still not finalized these limits. And several environmental experts argue that the proposed tighter standard itself does not go far enough. Nor do the proposed rules require that steps be taken to actually reduce lead levels from existing plumbing, even where problem pipes and fixtures have been identified. (Neither state nor city regulations pack much of a wallop in this area.)

Lead in drinking water is certainly not the major source of lead exposure for New York City children. However, there may be an additive effect from low-level exposure over time. Parents who fill up a bottle of water just after washing the dinner dishes for use the next morning are acting prudently to reduce risks to their children from unnecessary exposure to this toxic metal.

Lead in Air

What is the most significant environmental advance in New York City over the last 15 years? Some observers could suggest the ongoing multi-billion-dollar rehabilitation of the city's subway system, where gleaming new cars on some lines are the first signs of much-needed improvements in transit service. Others might point to restoration efforts in some of the city's most popular parks. Allow us to nominate the improvement in air quality brought about by a dramatic reduction of toxic lead in gasoline.

Since the 1920s lead has been blended into motor vehicle fuel to boost octane ratings and prevent engine knocking. Consequently, gasoline emissions have long been the largest source, by far, of airborne lead discharges. By the early 1970s, motor vehicle lead emissions just in the New York metropolitan area were well in excess of 3,000 tons a year. [13]

The struggle to tame airborne lead pollution in the city began with the passage of the 1970 Clean Air Act and the enactment of a 1971 city law providing for the gradual elimination of leaded gasoline sales within the five boroughs. In 1973, the U.S. Environmental Protection Agency adopted a nationwide lead-in-gasoline phasedown program. Court challenges, delays, and postponements followed one after another. Years slipped by. And the Reagan administration actually attempted to rescind the lead-in-gas phasedown rules entirely in 1982. But a sustained outcry from the scientific community and the public precipitated a 180-degree turnaround. And, by 1985, the amount of allowable lead in fuel had declined by more than 95 percent from uncontrolled levels in the early 1970s. [14]

The lead-in-gas phasedown triggered a nosedive in New York City airborne lead emissions. Gasoline lead discharges have been shaved to a fraction of what they were in the early 1970s. In fact, leaded grades of gasoline have actually vanished at many service stations around town. New York City now meets EPA's national air quality standard for lead. Although precise comparisons are difficult because of monitoring deficiencies, data indicate that lead levels of air in the New York metropolitan area have dropped by over 95 percent since 1970. This decline has been matched by dramatic fall-offs in blood lead levels of New York City preschool children. [15]

From a citywide perspective, this is great news. But there can still be localized problems from individual sources of airborne lead. Fortunately, there are not many industrial facilities emitting this toxic metal within the five boroughs. But there have been a few. The Non-Ferrous Processing Corporation, a lead-smelting plant in the Greenpoint section of Brooklyn, had long been the most troubling single source of industrial lead pollution in New York City. In 1980, its facility was emitting more than 40 tons of lead per year, at levels more than four times the national health standard. Under a consent agreement with the U.S. Environmental Protection Agency, Non-Ferrous agreed to install pollution controls to cut back these discharges. [16]

A significant remaining source of airborne lead in New York City

FIGURE 5.2: LEVELS OF LEAD IN GASOLINE AND NEW YORK AIR

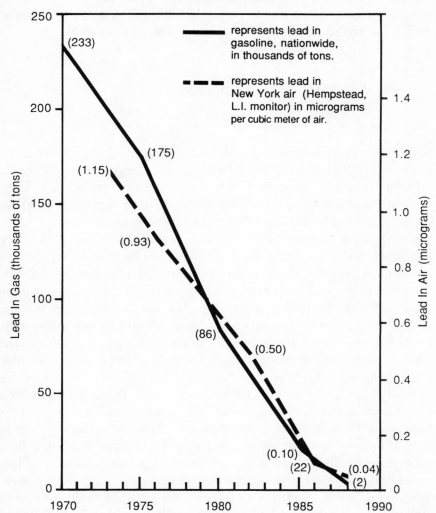

U.S. Environmental Protection Agency; New York State Department of Environmental Conservation

today are incinerators. The 2,200 apartment house incinerators, the scores of hospital incinerators, and the three existing municipal trash-burning plants continue to blow lead into city air. Little data are available on just how much. But the proposed 3,000-ton-a-day Brooklyn Navy Yard garbage-burning plant is expected to send 15 tons of lead up its stacks annually, according to city documents. If the city ultimately builds a series of such large trash incinerators, emissions from these plants may add a troubling footnote to New York's successful story of airborne lead pollution control.

ASBESTOS

Background

Sometimes environmental problems seem to appear right out of the blue. City officials have thought they were getting on top of the asbestos threat in New York. Over the last few years, they have begun to implement a comprehensive law that regulates asbestos removal during home and office renovation or demolition.

During the summer of 1989, a series of explosions and other incidents involving Con Edison's asbestos-lined steam pipes in Manhattan brought a new source of potential asbestos exposure to the front pages and the local newscasts. City agencies were caught off guard and scrambled to develop cleanup and decontamination procedures. The steam pipes problem should force utility and government experts to design preventive strategies, even as city officials press ahead on the larger issue of asbestos in homes and offices, the main focus of this section.

Most New Yorkers don't realize how much asbestos is around. This fiberlike substance has been used for much of the century in thousands of commercial products and building materials. It is nonflammable, heat resistant, and extremely durable. You'll find asbestos as insulation around pipes, boilers, radiator covers, and hot water tanks, and in textured wall and ceiling coverings. Even today, asbestos is being manufactured and used in brake and clutch linings, floor and roofing tiles, cements, and a variety of paper products (although its days may be numbered as a result of recent federal action).

Recognizing that asbestos can be a killer, labor leaders and government regulators have stepped in to make its control a top environmental priority in New York City. This is a ticklish assignment. The challenge is to remove or cover asbestos where it poses a threat to workers or the public, without making matters worse through unnecessary or haphazard abatement or illegal disposal. One can acknowledge that this dilemma exists, while still crediting the city for setting in motion one of the most comprehensive asbestos control programs in the nation.

Asbestos: Health Effects

Nobody would be going out on a limb in calling asbestos a highly toxic environmental contaminant. Studies have repeatedly linked asbestos inhalation in occupational settings to increased incidence of lung cancer and to a rarer form of cancer in the lining of the chest or abdomen. It has been estimated that between 3,300 and 12,000 cancer cases a year occur in the United States as a result of past exposure to asbestos. Asbestos exposure may also cause asbestosis, a serious respiratory disease in which lung tissue is permanently scarred. As is the case with other environmental pollutants, the effects of asbestos inhalation are compounded by cigarette smoking. [17]

Short-term, high-level exposures to asbestos, as well as long-term, low-level ones, are associated with cancer. As the National Institute for Occupational Safety and Health concluded more than 15 years ago, "The effect after several decades of a one-time acute dose of asbestos of limited duration which overwhelms the clearing mechanism, and is retained in the lungs, may be as harmful as the cumulative effect of lower daily doses of exposure over many years of work." Scientists agree that there is no known safe level of asbestos exposure. [18]

Asbestos Exposure in New York City

For many environmental problems, the first alarms resonate from the workplace. One example is asbestos. Increased deaths and disease in miners, shipbuilders, and factory workers exposed to asbestos confirmed the dangers of this toxic substance more than three decades ago. Now, the profile of workers in danger is changing. Today, it is abatement workers, building custodians, maintenance and utility personnel, who are at highest risk in New York

City. In fact, the removal or disturbance of asbestos by poorly trained or unprotected laborers may touch off a major increase in occupational exposures here in the coming decade.

The asbestos threat does not, however, stop at the occupational doorstep. Health experts warn of risks to the public. Between the 1950s and the 1970s, asbestos was routinely sprayed on or installed in New York City homes, offices, and buildings. As asbestos-containing material has deteriorated or been disturbed during building renovation, New Yorkers have been placed at risk from inhaling its airborne fibers. It is so-called friable asbestos—material that can easily crumble or have its fibers jarred loose—that is most susceptible to damage or deterioration. Once airborne, the microscopic fibers can remain suspended for many hours and may be kicked up again during dusting or vacuuming.

Asbestos emissions from motor vehicle brakes are another, yet often overlooked, source of public and occupational exposure. Mechanics who service brake equipment are at greatest risk. But asbestos released during routine driving and braking increases levels of this contaminant in air the public breathes, particularly in cities like New York, famous for its heavy volume of stop-and-go traffic. [19]

In New York City, the asbestos threat in commercial and residential buildings is not merely speculative. A 1988 citywide survey projected that about 500,000 buildings, or two-thirds of all such structures, have some form of asbestos, with 84 percent of the asbestos damaged in some way. Much of this asbestos is located in boiler rooms, heating ducts, elevator shafts, and other maintenance areas. [20]

How about the schools? A 1985 Board of Education survey of the city's approximately 1,000 public school buildings identified potentially troublesome asbestos-containing materials at 250 sites, including auditoriums, libraries, and sound rooms; asbestos was also discovered at 700 additional out-of-the-way locations, such as in boiler rooms. [21] Asbestos in schools is worrisome because of potentially extended periods of exposure for students and the long future lifetimes over which disease can develop. (The latency period, or time it can take before asbestos-induced health problems can occur, is 20 to 40 years.)

Asbestos: The Law

New Yorkers seeking to comprehend the federal approach to asbestos regulation must sift through at least four major laws. Federal
agencies implementing asbestos controls have at times relied upon
the Clean Air Act, the Toxic Substances Control Act, the Occupational Safety and Health Act, and legislation aimed specifically at
reducing classroom exposures.[22]

As is its method for dealing with many toxins, the federal government has taken the compromise route in regulating asbestos. Officials have acknowledged the danger of this material to public health
and have prohibited some uses. For example, they have placed the
application of virtually all asbestos in spray-on form and for pipe
insulation off limits. They have also gradually moved to protect
workers from breathing asbestos fibers. But go to your local hardware store and you may still find asbestos-containing products on
the shelf. To be sure, recently adopted EPA rules will phase out
94 percent of current asbestos uses, but not until 1996.[23] And federal officials have sidestepped the difficult question of how to reduce public exposure to asbestos materials and products still
in use.

There is one exception to the general reluctance of Washington
decision-makers to confront the problem of in-use asbestos exposure outside of the workplace. EPA has recently sought to reduce
the asbestos threat in the nation's schools. Agency rules direct local
school systems to survey their facilities and draw up abatement
plans.

Only recently has the state legislature waded into the regulatory
waters on the asbestos front. To help curb fly-by-night asbestos
removal, the legislature in 1986 required the licensing of asbestos
abatement contractors and the training and certification of asbestos workers. It also directed the Labor Department to adopt regulations governing the removal or encapsulation of asbestos
material.[24]

State attorney general Robert Abrams, a frequent friend of the
environment, has used his own authority to protect consumers
from asbestos exposure. In 1986, he adopted regulations applicable
to buildings undergoing co-op or condominium conversions. The

portion of these rules that survived court challenge requires co-op and condominium sponsors to inspect apartments for the presence of asbestos and to disclose the findings to residents and potential purchasers.[25]

New York City lawmakers first attacked the asbestos problem in the early 1970s and then took a 13-year leave of absence. In 1972, the City Council banned the application of spray-on asbestos in building construction. But it was not until 1985 that city legislators began to confront the nagging problem of in-place asbestos. They enacted new legislation to prevent the release of asbestos during demolition or renovation activities. The city law, like state legislation that followed it, also required training and certification of asbestos removal workers. Another City Council enactment passed that same year governs the storage, transport, and disposal practices of asbestos, helping to complete the statutory loop.[26]

Asbestos: Government Action

In New York City's war against asbestos, government agencies have begun homing in on the enemy's main launching pads—the thousands of buildings where demolition or renovation activities threaten to discharge asbestos into the environment. Government's principal weapon has been the 1985 city law that arms enforcement officers with regulations covering almost every aspect of the asbestos removal business. The regulations are detailed. More than 200 separate provisions cover everything from restrictions on public access at work sites to requirements for the safe removal, isolation, and cleanup of asbestos wastes.[27]

A team at the Department of Environmental Protection (DEP) has drawn the assignment of implementing the asbestos control program. Their results have been mixed. During a typical month in 1989, approximately 500 to 600 persons notified the city, as required by law, that they were undertaking major demolition or renovation work in which asbestos might be disturbed. Over that same period, DEP investigators were making inspections at roughly 200 sites (in some cases more than once) and were issuing approximately 50 to 60 violations. Department officials report that from mid-1987 (when the rules took effect) to mid-1989, they have

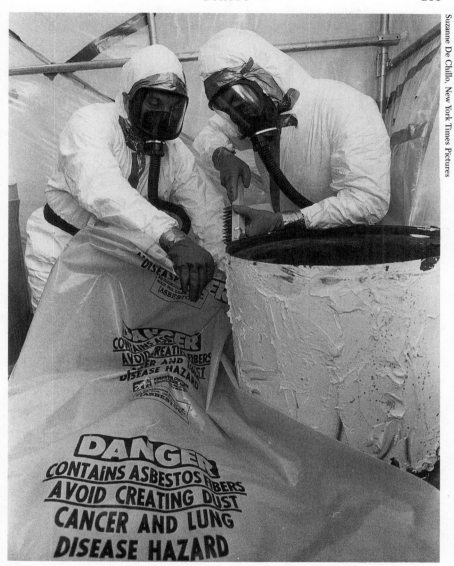

Suzanne De Chillo, New York Times Pictures

THE ASBESTOS DILEMMA. *Asbestos, a known carcinogen, is present in roughly two-thirds of all buildings in New York City. Some experts caution that while friable, or crumbling, asbestos should be properly removed by certified workers, undamaged asbestos may be better off left in place and safely covered. The biggest risk is to asbestos removal workers, custodians, janitors, and others exposed in occupational settings.*

seen improvements in compliance, with fewer gross violations of the regulation's protective umbrella.

But the city's tough asbestos law is only as strong as its enforcement punch. And DEP's 50-person asbestos unit had only 12 inspectors as of May 1989. As a result, the department's investigators were simply unable to visit more than half of the sites where building owners reported to DEP that asbestos was being disturbed.[28]

The need for aggressive enforcement on the asbestos front could not be clearer. In early 1988, officials at 23 companies—reportedly representing a majority of the firms handling asbestos removal in New York City at that time—were indicted and charged with bribing a U.S. Environmental Protection Agency inspector to overlook violations of federal asbestos rules. Among these contractors was the Bronx-based Big Apple Wrecking Corporation, one of the city's largest demolition companies. Its president was subsequently convicted for bribing the federal inspector at the old Gimbel's department store sites at Herald Square and on Manhattan's Upper East Side.[29]

Also troubling is the unknown number of asbestos removals that are taking place without any notification to the city and presumably without environmental safeguards. Some insiders call these unlawful removals "rip and skip." The incentive is financial—it's cheaper to simply tear out the asbestos and dispose of it unlawfully.

New Yorkers who are living in small buildings or private homes in which asbestos is damaged face the greatest hurdles. They are often uneducated about asbestos and its dangers. They lack technical information on proper encapsulation or removal. And they have few places to turn for help (although the nonprofit White Lung Association fills part of this need). Many of their properties are not covered by the city's rules because of the relatively small quantities of asbestos present. Most city regulations governing asbestos removal do not apply to properties of this size. And even homeowners or small property owners who actually seek professional asbestos removal services find it difficult to secure them. Most asbestos contractors prefer the larger jobs; their services are expensive and no financial assistance programs are available.

Existing city law looks the other way when it comes to damaged asbestos in buildings not undergoing renovation or demolition. This is a major gap that city lawmakers and administration officials are attempting to close in 1990. Their proposed legislation (Intro. #1164) would require that building owners inspect for asbestos and prepare plans to cover, remove, or maintain in-place asbestos, as necessary, over the next 6 to 12 years. (Two-family and smaller homes would be covered only upon title transfer.) The bill would also clarify the city's authority to inspect for damaged asbestos and order cleanup, regardless of whether buildings were undergoing renovation work. Should this proposal be enacted into law, it is likely that some property owners will race the statutory time clock and rip out asbestos willy-nilly before the new requirements take effect. Despite such problems, passage of the bill would be an overall plus.

A final note on the city schools, where media attention on the asbestos problem has frequently been aimed. Under federal regulations, the New York City Board of Education has recently drawn up asbestos abatement or maintenance plans for all school buildings. School officials, who see at least ten more years of cleanup work ahead, have generally been tight-lipped about specific conditions at individual schools. But, fair or not, how well the Board of Education carries out these duties will persist as a popular measuring stick on asbestos control in New York City.

PCBs

Background

"Without chemicals, life itself would be impossible," an industry advertising campaign proclaimed a while back. The slogan writers were onto something. They recognized that our society has in many ways become dependent on chemicals. And they reminded the public that some advances in medical science have come about thanks to the nation's chemical and pharmaceutical companies.

But there's another side to the story that the industry ads ignored. The country's chemical courtship has come with significant costs. For example, the environmental and public health threats from a handful of chemicals have become so compelling that government officials have actually banned their further production. One member of this infamous club is polychlorinated biphenyls, or PCBs. But years after government restrictions on most PCB production, there remains a troubling postscript—the impacts of this toxin continue to linger on the New York environmental scene.

PCBs: Health and Environmental Effects

First, a quick profile. PCBs are synthetic, organic chemicals. In liquid form, they have a heavy, oil-like consistency. PCBs have been used for decades as insulating or cooling fluids in electrical equipment like transformers and capacitors because of their heat-resistant properties and low conductivity. They have also been mixed into paints, adhesives, hydraulic fluids, and dyes.[30]

PCBs are major league environmental contaminants. There is less information on the human health impacts of PCBs than other long-studied pollutants like lead or asbestos. But scientists do know that PCB exposure can cause chloracne (a painful and disfiguring skin condition), liver damage, abdominal pain, digestive problems, and nervous system disorders. PCBs are also suspected of causing birth defects and are a probable human carcinogen. When released from transformers during electrical fires, PCBs can turn into gaseous form, and may be converted into even more hazardous compounds of dioxin. PCBs are toxic to fish and other animal life at very low levels of exposure. Some forms of this toxin are among the most stable chemicals known; their danger to marine life is compounded because they persist in the environment for years and can accumulate in the food chain.[31]

PCB Exposures in New York

When toxic chemicals are in circulation, unexpected problems can pop up. Look no further than the two primary PCB exposure threats to New Yorkers.

First, there's the GE saga. Few people would have guessed that in the late 1940s, when the General Electric Company started discharging by-products from electrical equipment manufacturing into the Hudson River, it would turn into one of the state's most nettlesome environmental problems. But that is exactly what has happened. From 1947 through the mid-1970s, two GE capacitor plants north of Albany sent more than 500,000 pounds of PCBs into the river. Most of the toxins were deposited along a 40-mile stretch, but some contaminants were later carried as far south as New York Harbor and the rich spawning grounds of the Hudson River estuary.[32]

If you're eating fish from the Hudson River, there's a good chance you're getting a side order of PCBs. State and federal environmental agencies have called the Hudson "perhaps the world's most severely PCB-contaminated aquatic ecosystem." PCBs have been detected in more than 20 species of fish taken from this river and surrounding New York waters. In recent years, PCBs in the river's renowned striped bass have sometimes been found at levels two or more times federal standards for safe consumption. For these reasons, state officials have placed commercial fishing for nine species off limits and have issued health warnings on consumption of more than a dozen species to recreational fishing enthusiasts. Combined economic losses to the region's commercial and recreational fishing industries from PCB contamination have been estimated to be as high as $38 million a year.[33]

PCBs in electrical equipment pose a different problem. For most New Yorkers, risks are minimal so long as the compounds remain fully enclosed. But PCBs can escape through fires, explosions, or other accidental releases, with eye-opening consequences. In 1981, an electrical fire broke out in a New York State office building in the upstate community of Binghamton, spreading PCBs and dioxins throughout the 18-story tower and closing that city's tallest building ever since. The total price tag for toxic cleanup is projected to be $40 million—more than double the original cost of construction.[34]

Can a Binghamton-type fire occur in New York City? Possibly. In 1987, there were roughly 1,440 large PCB transformers in use in the city's five boroughs, according to Fire Department records. And from January 1988 to January 1989 there were 15 PCB spills and one minor fire involving PCB electrical equipment, say federal officials who track such events.[35] Although very little information on PCB exposure from these incidents or others is available, New Yorkers seem so far to have escaped unscathed from any sudden, high-exposure PCB release.

CHEMICALS IN YOUR COMMUNITY

Does it do any good to let citizens know that there are toxic chemicals being stored and even released in their neighborhood?

Congress thought so in 1987, when it passed community right-to-know legislation in the wake of the Bhopal, India, chemical disaster. One prong of the law requires certain industries to submit information to government authorities on chemical releases into air, land, or water. The other mandates that companies disclose what chemicals they are storing on-site. It also directs localities to develop emergency response plans.[37]

New York City shared the congressional enthusiasm for information gathering. In 1988, it enacted a local right-to-know law that effectively expands the number of facilities and the number of chemicals subject to toxic chemical disclosure requirements.[38]

Data that most industries have preferred not to disclose are now making their way into the public domain. Under the statutes, several hundred New York City companies have filed disclosure forms with government authorities. The information being reported is perhaps most appreciated by Fire Department officials and has also stimulated at least some emergency planning by an interagency city task force.[39] Of course, public disclosure isn't the same as pollution control. But, as the saying goes, knowledge is power.

PCBs: The Law and Government Action

Of the more than 60,000 chemicals now in production, PCBs were the only ones singled out by name when Congress passed the Toxic Substances Control Act in 1976. That law authorized EPA to prohibit (with limited exceptions) the manufacture, distribution, and use of this chemical. In 1979, EPA promulgated regulations that restricted PCB use, except for PCBs in enclosed systems and a handful of other products. The Binghamton fire and similar incidents triggered an EPA reassessment. And, in 1985, the agency banned the installation of new PCB transformers to be located in or near commercial buildings. It also prohibited the use of some existing transformers in such locations, effective 1990. In response to such edicts and in the hope of avoiding a costly mishap, city agencies, Con Edison, and others are draining some of their equipment of PCBs or removing the equipment completely.[36]

The New York State thrust on PCBs has centered on decontaminating hotspots in the Hudson River. It has been a most frustrating venture. For more than a decade, state officials have been searching in vain to find an acceptable location for land burial of the unwelcome sediments. Two different sites in the upstate Washington County community of Fort Edward have been tapped at one time or another by Environmental Conservation planners as a final resting place for the PCBs. But local opponents and others have succeeded in blocking the toxic transfer. There seems no easy way out of this logjam.

HAZARDOUS WASTES

Background

Some government officials would have you believe that there is no hazardous waste problem in New York City. No Love Canals, they say. No mammoth industrial facilities. And an apparent fall-off in illegal dumping, at least at city landfills. There's a grain of truth to this argument.

But hazardous wastes are a problem in New York City, even if the story has not been front-page news. Without much fanfare, industries, small businesses, and households within the five boroughs have been generating tens of thousands of tons of hazardous waste every year. Much of this material has been disposed of informally—in landfills, into sewers, and through other loosely regulated disposal methods. Past practices have left dozens of hazardous waste disposal sites scattered throughout the city. And a handful of these sites have made their way onto the state's priority list of locations posing a significant threat to health or the environment.

In many parts of the country, the hazardous waste issue has served as a focal point for environmental activism. This has not occurred in New York, in part because there has been no single

HOUSEHOLD HAZARDOUS WASTE

It's easy to pin the hazardous waste problem exclusively on the lapel of industry. But don't forget households when tallying the city's hazardous waste generators. New York's garbage contains dozens of consumer products that, when disposed of, may unleash their toxic properties.

How many items on this toxics inventory can be found in your home?

COMMON HOUSEHOLD TOXICS

Kitchen
Roach and ant killers
Silver polish
Oven cleaner
Floor or furniture polish

Bathroom
Toilet cleaner
Disinfectants
Drain cleaner
Ammonia-based cleaners
Photography chemicals

Bedroom
Furniture strippers
House plant insecticides
Rug and upholstery cleaner
Mothballs

Garage
Pesticides
Flea collars and sprays
Rat and mouse poison
Herbicides
Enamel or oil-based paints
Rust paints
Wood stains
Thinners and turpentine
Pool chemicals
Antifreeze
Transmission and brake fluid
Used motor oils
Car batteries

Most of these materials would, upon disposal, be formally branded "hazardous waste," if toxicity alone were the determining factor. But since they are consumer products and are disposed of by homeowners, they have a virtual immunity from the rules that apply to officially designated hazardous waste.

They are treated instead as ordinary household trash. So they usually wind up at the Fresh Kills landfill. Or they are burned in a municipal or apartment house garbage incinerator. Or they are poured down the drain and flushed into the sewer system. But city landfills leak. Existing incinerators cannot completely destroy such toxic concoctions. And sewage plants are unequipped to treat most toxic wastes. Some government estimates suggest that household toxics range from 1/3 to 1/2 of 1 percent of all residential waste. We are talking, then, about the disposal of perhaps 100,000 pounds a day of such materials in New York City.[42]

There is an alternative. It's called waste reduction. And it involves substituting less harmful items for the more toxic products New Yorkers usually buy from habit. For example, you can use boric acid or nontoxic traps instead of household pesticide sprays for controlling roaches and other insects.

But for some other products—like automobile engine oils or photographic chemicals—there are no substitutes. To dispose of household wastes like these, hundreds of communities around the nation are inaugurating household toxic collection days. The idea is to segregate these wastes from ordinary trash—recycling them when possible and disposing of the remainder in a way similar to that for industrial hazardous waste.

City officials aren't racing to schedule regular household toxic waste collection days in the five boroughs. So far, they have only tiptoed into a one-time demonstration. And with many environmental programs competing for scarce city resources, things here are unlikely to change any time soon.

corporate black sheep responsible for a sizable piece of the city's hazardous waste problem. But the lack of political militancy here does not remove the need to deal with existing environmental harm. Hazardous waste may not be the city's first environmental priority, but it certainly should not be its last.

Hazardous Waste Generation

How much hazardous waste is generated in New York City? No one really knows. For one thing, there is no clear definition of what constitutes hazardous waste. Many chemical waste products that may be dangerous to human health or the environment are not included in official government hazardous waste production figures. Used motor oils, industrial wastewaters, radioactive wastes, and hazardous household products, for example, are not counted in this somewhat arbitrary classification scheme. To make matters worse, a comprehensive assessment of waste generation has been hampered by gaps in industry reporting and data collection. No citywide analysis has ever been completed. It appears, however, that total production levels are well in excess of 50,000 tons a year. [40]

Hazardous waste generation in New York City is surprising in at least one major respect. In some localities, a few dominant facilities produce the bulk of the region's toxic waste. But, in New York, it is the small and mid-size hazardous waste producers that represent a significant chunk of the hazardous waste problem. Perhaps as many as 13,000 may be operating in New York City. Gasoline service stations, motor vehicle repair shops, dry cleaners, electroplating and metal manufacturers, photo labs, and printing and dyeing operations are among the wide-ranging generators. Wastes from these locations include lead acid batteries, cleaning and degreasing solvents, heavy metal sludges, paints and inks.

Of course, New York City is also home to some big-time hazardous waste generators. Official state figures put the number at about 70. Each of these producers churns out more than the equivalent of 60 drums of hazardous waste a year. The two largest generators by far are Consolidated Edison and the New York City Transit Authority. [41]

Hazardous Waste Disposal

The shadowy process of hazardous waste disposal in New York City has gone on largely outside the public limelight. Like the nation-wide experience, the disposal of hazardous waste in New York has until recently occurred without significant government scrutiny. Some industries have surreptitiously sent their waste materials to city-owned landfills. Others have stockpiled wastes indefinitely on their premises, creating *de facto* disposal sites. And not infrequently, industrial wastes have simply been abandoned on vacant city lots.

The New York State Department of Environmental Conservation has currently placed 29 city locations on its list of inactive hazardous waste disposal sites requiring special attention. And they have designated nine locations as ones that pose "a significant threat to public health or the environment." The City Sanitation Department is the not-so-proud landlord of five of the largest and perhaps most troubling of these sites, all former garbage dumps, where illegal industrial dumping apparently went unchecked for years. And it is likely that the total number of problem sites could be larger, according to reconnaissance by federal investigators.[43] Federal officials have not moved any New York City site onto their highly competitive national Superfund list, a reflection of how serious the situation is in other parts of the country.

The adverse environmental and public health risks from past disposal practices vary from site to site and have not been the subject of full-scale investigation. But indications are that the primary danger from many of these sites is contamination of city waterways. At Brooklyn's Pennsylvania, Fountain, and Edgemere landfills, for example, leaching toxic metals and organic chemicals threaten the natural resources of Jamaica Bay. Or take the Quanta Resources Corporation site, a former waste oil reprocessing facility in Long Island City. PCB-contaminated oils that had been collected in on-site storage tanks at the now bankrupt corporation may be migrating, through the underground water table, into nearby Newtown Creek.

To be sure, the days of widespread, unbridled hazardous waste disposal may be numbered. Some generators are transporting their

TABLE 5.1 TWENTY-NINE INACTIVE HAZARDOUS WASTE DISPOSAL SITES IN NEW YORK CITY

Site	Location	Classification
Pelham Bay landfill	Bronx	2
MTA Gun Hill Bus Complex and adjacent lots	Bronx	2a
Hexagon Laboratories	Bronx	2a
Pennsylvania Avenue landfill	Brooklyn	2
Fountain Avenue landfill	Brooklyn	2
Hamilton Avenue piers/19th and 18th streets	Brooklyn	2a
Silver Rod Drug Company	Brooklyn	2a
Rear of Bush Terminal Building	Brooklyn	2a
Carroll Gardens	Brooklyn	2a
Spring Creek (Emerald Street)	Brooklyn	2a
Empire Electric Company	Brooklyn	2
BQE/Ansbacher Color and Dye Factory	Brooklyn	2a
College Point Oil Lagoon	Queens	4
Phelps Dodge Refining Corporation	Queens	2
Edgemere landfill	Queens	2
Quanta Resources	Queens	2a
Amtrak Sunnyside Yard	Queens	2
DEKNATEL (Division of Pfizer Hospital Products)	Queens	2a
Idlewild Construction Waste landfill	Queens	2a
Levco Metals Property	Queens	2a
Positive Chemical	Staten Island	5
Vigliarolo Bros. (Onyx Chemical Company)	Staten Island	3
Brookfield Avenue landfill	Staten Island	2
Arden Heights Shopping Mall	Staten Island	3
R. T. Baker & Son Machinery Salvage Company	Staten Island	2a
Arden Woods Estates, Inc.	Staten Island	2
A&A landfill	Staten Island	2a
Pergament Mall/Corniche Dry Cleaners	Staten Island	2a
Allied Prince's Bay	Staten Island	2

Classification 1: Causing or presenting an imminent danger of causing irreversible or irreparable damage to the public health or the environment—immediate action is required.

Classification 2: Significant threat to the public health or environment—action required.

Classification 2a: Temporary classification of sites for which insufficient data are available.

Classification 3: Does not present a significant threat to the environment—action may be deferred.

Classification 4: Site properly closed—requires continued management.

Classification 5: Site properly closed, no evidence of present or potential adverse impact—no further action required.

New York State Department of Environmental Conservation

wastes to landfills or incinerators outside the city, which are officially designated to receive these materials. Others, such as some
service stations and dry cleaners, have begun to recycle a portion of
their wastes (e.g., automotive batteries or used cleaning solvents,
respectively). But despite such indications, it is difficult to gauge
how much illegal or improper disposal is continuing. And even
where disposal practices are in compliance with existing regulatory
requirements, the long-term environmental effects of some disposal methods are still unknown.

The Law: Hazardous Wastes

Congress has ventured into the area of hazardous wastes through
two major enactments. The 1976 Resource Conservation and Recovery Act (RCRA) was the first such excursion. It directed EPA to
lay out a comprehensive framework for insuring that hazardous
waste is properly handled from generation to ultimate disposal.
Under this mandate, EPA has, among other things, defined which
wastes are hazardous, established conditions and a tracking ("manifest") system for waste shipment, and set minimum standards for
waste treatment, storage, and disposal facilities.

Congress authorized EPA to turn the reins of the RCRA program
over to the states, once the agency has determined that a state's
regulatory scheme is at least as stringent as federal law. In response,
New York State enacted its own hazardous waste control program in
1978, and received final federal approval in 1986.[44]

Congress next took up the gauntlet with the 1980 passage of the
federal Superfund law. This statute for the first time made federal
monies available for emergency and long-term cleanup of the nation's worst hazardous waste disposal sites. The cleanup program
(formally known as the Comprehensive Environmental Response,
Compensation and Liability Act, or CERCLA) and its 1986 amendments (the Superfund Amendments and Reauthorization Act, or
SARA) are funded by taxes on petroleum products and business
income, and by state matching funds. Under the federal law, Superfund monies can only be allocated to the long-term cleanup of sites
that have officially been included on a national priority list adopted
by EPA.[45]

State and federal environmental laws often work in tandem. New York's 1982 state Superfund law is illustrative. It provides state matching funds needed to secure federal cleanup monies. It also supplies additional funding for New York cleanup operations at hazardous waste sites not included on the federal Superfund list. To set the state cleanup troops in motion, the law requires the Departments of Environmental Conservation and Health to evaluate potential hazardous waste sites and rank them for remedial action.[46]

Government Action: Hazardous Waste

What impact is the complex web of government regulations having on day-to-day handling and disposal of hazardous waste in New York City? There are few indicators on which to base any firm conclusions. Little hard data on the number of hazardous waste producers, particularly small-quantity generators, have been collected. Nor have government officials gotten a handle on how much of these wastes are produced in the first place. And the involvement of middlemen who often transport or store hazardous materials has complicated government monitoring. About all that can be said is that the long and arduous process of hazardous waste record keeping, permitting, and tracking has finally been set in motion.

On the cleanup front, state authorities have taken the lead in overseeing work at the city's inactive hazardous waste disposal locations. Dozens of suspected sites have been surveyed in recent years. But progress in actually cleaning up New York City toxic waste dumps on the state Superfund list has been sluggish. Small-scale cleanups at some sites have been completed. A one-half-acre lagoon containing roughly 500,000 gallons of PCB-contaminated oil and wastewaters in College Point, Queens, for example, has been cleaned and filled by the city. But remediation at most of the larger sites has not significantly progressed.[47] This is the case at nearly all of the highest-ranking sites on the state hazardous waste registry.

There are real questions as to whether the city will ever be able to cleanse its inactive hazardous waste sites in their entirety. Local efforts in New York (and elsewhere) have largely been limited to aboveground cleanup. At the recently closed Pennsylvania Avenue landfill, for example, city engineers installed an absorbent boom in surrounding waters to impede the flow of contaminated oils from entering Jamaica Bay. But millions of gallons of waste oils and other toxic substances remain buried beneath the surface. The task of full-scale subsurface cleanup at many sites is considerable and government officials have no present plans to undertake such efforts. Cleanup of hazardous waste sites in New York City, then, is sliding into a program of waste containment, rather than full-scale waste removal.

TOXICS IN THE WORKPLACE

Background

Relations between the labor movement and the environmental community have sometimes been tense. Some in the labor camp have feared that the environmental push would shut down businesses and eliminate jobs. The situation has proven otherwise. On balance, environmental laws have created many more new jobs (building sewage treatment plants, manufacturing pollution controls, etc.) than have been lost from the isolated closure of marginal plants. Nevertheless, some animosity has lingered.

Despite occasional flare-ups, labor and environmental activists have been drawn together by common concerns in the area of occupational health. These two natural allies have recently coalesced around such issues as pesticide law reform and nuclear weapons plant safety. And a détente between these factions may now be emerging on the New York City scene.

Concern over exposure to toxic substances is what has brought on the handshaking here. Both groups agree that the highest levels of exposure for many toxins can be found in the workplace. There are few comprehensive surveys that quantify the scope of this problem in New York. But the Mount Sinai School of Medicine, a national leader in the diagnosis and treatment of occupational disease, has reported that an estimated 5,000 to 7,000 deaths are occurring across the state each year from work-related exposures.[48]

To understand the occupational problem, one must look beyond the toxicity of workplace chemicals alone. For one thing, there's an information gap. Some workers are unaware of the on-the-job environmental dangers they face. What's more, occupational diseases often go unrecognized; many health care professionals still lack adequate training to diagnose these illnesses. And even when the risks are well known, some employees are reluctant to bring their concerns to management for fear of retaliation. Finally, government agencies have been slow to order their troops into the fight over workplace toxins. (The government's battles, if you want to call them that, have largely been fought over traditional issues of worker accidents and safety, which, while important, are not the focus of this discussion.)

Progress in curbing occupational diseases in New York City won't come overnight. Major new thrusts by federal, state, and city agencies in this area are likely to trail high-visibility calls for action, as was the case with asbestos. Further gains on the occupational front then will require every bit of coalition building that labor groups and their supportive constituencies, including environmentalists, can mobilize.

Occupational Exposures

People will tell you that every job has its risks. Many employees have traditionally accepted workplace dangers as part of the job. And despite a growing recognition that chemical exposures can cause occupational disease, there remains a reluctance in some quarters to tie employee illnesses to the workplace. One explanation is that many occupational diseases can take decades to appear (making cause-and-effect relationships difficult to prove). Another is that such personal factors as age, smoking, diet, and stress can influence the onset or severity of occupational disease (so that responses vary, even among co-workers who receive similar exposures to the same chemical agent). Additionally, proving the linkage between disease and workplace exposure is complicated by the combination of toxins that can be found on the job. According to one federal survey, the number of potentially hazardous chemical substances to which U.S. workers (as a whole) may be exposed exceeds 8,000.[49]

TOXIC TINDERBOX?

If hazardous waste management programs are working in New York City, it is not obvious to some residents of Williamsburg, Brooklyn. Local citizens there are frustrated. Despite stacks of federal, state, and local rules governing waste management, a private company has recently secured a state permit to continue storing toxic and potentially explosive chemicals in a densely populated neighborhood, ill suited for such a facility.

The Radiac Research Corpora-

tion is a transfer and storage operation for hazardous and chemical wastes. Tens of thousands of pounds or more of such materials are stockpiled at this innocuously named facility until their 30- or 55-gallon drums are hauled away, usually to out-of-state disposal sites. Only 30 to 35 percent of these wastes comes from New York City and New York State; some waste is trucked in from as far away as Massachusetts. First opened in this building in 1976, Radiac is appar-

Eric A. Goldstein

RADIAC RESEARCH CORPORATION, *Williamsburg, Brooklyn.*

ently one of only three such operations holding a waste storage and
treatment permit in New York
City. Many hazardous products
are housed at the Radiac facility,
located on South First Street, a
few blocks from the East River.
Among them are flammables (such
as acetone and other chemicals
with very low flash points), poisons
(such as arsenic), pesticides (including DDT), corrosives, acute
toxins (like cyanide), and reactives
(such as methyl isocynate—the infamous Bhopal, India, chemical).

What has angered commu-
nity groups like Williamsburg-
Around-the-Bridge Block Association and El Puente is the danger of
fire, explosion, or leakage from the
collection of toxins tucked away in
Radiac's one-story building. Hundreds of drums of incompatible

chemicals are stacked in a single
35-by-100-foot room. Low-level
radioactive wastes are housed in
an adjacent storage building, also
owned by Radiac. A metal-
welding operation is located just
next door. And a 1,100-student
public elementary school, several
food-processing facilities, and a
number of residential buildings sit
incongruously within a one-block
radius. According to one expert
who testified at Radiac's recent
permit hearing, these circumstances are creating "an invitation
to disaster."[50]

How could state officials allow
such a facility to be sited at this location? Isn't there anything that
prohibits such dangerous operations so close to so many incompatible land uses? Part of the answer is
that the Radiac operation was in

What are the high-risk jobs in New York City? Keep in mind that
we are talking only about dangers from toxins here, not workplace
accidents or injuries. You won't find complete agreement in this
area. And government agencies don't seem to have such a checklist.
But piecing together observations from experts in the field and
scientific case studies, it is possible to come up with a dozen or so
candidates.

Although the city's economy is increasingly white-collar, the most
hazardous occupations here are not. The high-risk list includes such
blue-collar trades as construction, roofing, garment, metal, and
printing. And it includes exterminators, auto repair workers, and
service station attendants. Segments of New York's large municipal
work force are also well represented—firefighters; bridge and tun-

existence before the federal and state hazardous waste laws were fully implemented; for this and other reasons no environmental impact statement has ever been prepared for this waste warehouse. Also, a federal directive that such facilities be surrounded by a 50-foot buffer zone was waived by the state in a controversial 1988 ruling. State officials, propelled at least in part by pressure to keep such depots available to waste generators, accepted a Radiac claim that the building's brick walls obviated the need for a buffer zone.

The situation may not be much better at the other hazardous waste storage facilities in New York City. But there is little information on the safety record of the two remaining facilities, both in Queens—the Chemical Waste Disposal Corporation in Astoria, and Safety Kleen Corporation in Woodside.

Back at Radiac, government officials quietly concede that such a hazardous waste facility doesn't belong where it now stands. Publicly, it's a different story. "The department's job is not to deny permits, but to process them," one state environmental conservation attorney admitted, noting that New York needs the services Radiac provides. His comment suggests that officials would rather avoid the political difficulties of siting such a facility at a more appropriate location. But to the Williamsburg community, it seems that such outside considerations are bending the hazardous waste rules out of shape.

nel workers; and transit maintenance, sewage treatment plant, landfill, and incinerator employees. Hospital and laboratory workers, dry cleaners, and asbestos abatement workers round out the somewhat subjective listing (see table 5.2).

Several experts have proposed adding office workers to this short list. Although not traditionally thought of as a high-risk workplace, office buildings are receiving increased surveillance from occupational watchdogs. The potential dangers include passive cigarette smoking and indoor air pollution, especially in newer, tightly sealed buildings. But what is getting some people fired up these days is concern about video display terminals (VDTs). The highly respected trade journal *Microwave News* was among the first to report that programmers and other office workers who use VDTs

Matthew Borkoski/Folio

Jamie Phillips/Folio

Vic Delucia, New York Times Pictures

Neal Boenzi, New York Times Pictures

TOXICS ON THE JOB. *Perhaps the greatest toxic threats to New Yorkers lurk in occupational settings. As many as 5,000 to 7,000 deaths a year in New York State may result from workplace disease. Among the high-risk occupations in New York City, pictured above clockwise from the upper right, are roofing workers, auto repair workers, garment workers, and firefighters.*

for lengthy periods may be at increased risk from low-level electromagnetic radiation.[51] Although the evidence is not yet conclusive,
the potential population that may be at risk is high. The issue
deserves close watching.

The Law: Occupational Toxins

The landmark Occupational Safety and Health Act (OSHA) sought
to make up for lost time. It was the "grim history of our failure to
heed the occupational health needs of our workers," Congress said,
that led to the act's passage in 1970. Among other things, the law
empowers the federal Labor Department to adopt health standards
and dictate workplace practices limiting occupational exposures to
certain toxins. It also directs the department to inspect work sites
and, where necessary, penalize businesses to insure compliance
with the regulatory mandate. Only private industry is covered
under the federal OSHA statute. But a New York State law has
deputized state Labor Department inspectors to enforce federal
OSHA standards in state and local government workplaces.[52]

Prevention has become the hot word in occupational circles.
That's why expectations have been raised over worker right-to-
know laws enacted during the 1980s. In general, these programs
help lift the veil of secrecy on what chemicals are being used in the
workplace. This can be done in several ways—product labeling,
distribution of fact sheets on workplace toxins, and employee education and training programs. A New York State worker right-to-
know program was carved into law in 1980; the less comprehensive
federal program, which had been anticipated for more than a decade, did not take full effect until 1987.[53]

Government Action: Occupational Toxins

The story of the Pymm Thermometer Corporation offers as good an
example as any of the frustrating government response in the battle
for occupational health. William and Edward Pymm had for years
manufactured thermometers in their one-story factory in the Bushwick section of Brooklyn. Since the early 1970s, workers had been
exposed to toxic mercury vapors of varying intensities, both on the

TABLE 5.2 THIRTEEN OCCUPATIONS AT HIGH RISK FROM TOXIC HAZARDS IN NEW YORK CITY

Occupation	Estimated Number of Potentially Exposed Workers[a]	Selected Toxic Exposures[b]	Known or Suspected Health Effects[c]
1. Asbestos abatement workers	16,100	Asbestos	Asbestosis, lung cancer, cancer of the chest lining or abdomen, other cancers
2. Auto repair workers, service station attendants	29,100	Benzene and other evaporative hydrocarbons (from gasoline, auto body painting)	Nervous system disorders, leukemia
		Asbestos (from brake linings)	Asbestosis, lung cancer, cancer of the chest lining or abdomen, other cancers
		Lead (from radiator repairs)	Nervous system disorders, kidney disease and hypertension
3. Bridge and tunnel workers (toll booth and maintenance workers)	2,000	Carbon monoxide, diesel particulates, and other motor vehicle pollutants	Cardiovascular problems, chronic respiratory diseases, lung cancer
4. Construction workers (including painters and roofing workers)	89,800	Asbestos (from fireproofing, insulation and asbestos cement)	Asbestosis, lung cancer, cancer of the chest lining or abdomen, other cancers
		Lead (from metal welding and cutting)	Nervous system disorders, kidney disease, and hypertension

TABLE 5.2 (Continued)

Occupation	Estimated Number of Potentially Exposed Workers[a]	Selected Toxic Exposures[b]	Known or Suspected Health Effects[c]
		Solvents (from paints, paint thinners, paint strippers)	Dizziness, headache, nausea, eye and nose irritation, nervous system disorders, cancer
		Fiberglass	Chronic respiratory disease, lung cancer
		Coal tar volatiles (from roofing work)	Skin irritation, skin and lung cancer
5. Dry cleaners	6,000	Tetrachloroethylene, trichloroethylene (and other cleaning solvents)	Nervous system disorders, liver cancer, leukemia
6. Exterminators	3,800 (New York City licenses only)	Pesticides	Skin irritation, nervous system disorders, possible carcinogens
7. Firefighters	12,500	Carbon monoxide	Cardiovascular and respiratory disease
		PCBs, dioxins (from electrical equipment fires), and PVC	Cancer

TABLE 5.2 (*Continued*)

	Occupation	Estimated Number of Potentially Exposed Workers[a]	Selected Toxic Exposures[b]	Known or Suspected Health Effects[c]
8.	Garment and textile workers	68,900	Formaldehyde (from permanent press fabric)	Skin rashes, eye irritation, other respiratory problems, asthma, intestinal and kidney problems, possible carcinogen
			Benzene (from waterproofing solvents)	Leukemia
			Cleaning solvents	Nervous system disorders
9.	Hospital and health care workers; laboratory workers	224,300	Anesthetic gases	Reduced performance levels, spontaneous abortions, birth defects, probable carcinogens
			Ethylene oxide (from sterilization processes)	Skin, eye, and lung irritation, nervous system disorders, reproductive problems, possible carcinogen
			Chemotherapeutic chemicals	Mutagenic effects, cancer
10.	Metal workers; electroplaters	23,500	Lead and cadmium (from welding and cutting)	Nervous system disorders, lung disease, kidney disease, and hypertension

TABLE 5.2 (Continued)

Occupation	Estimated Number of Potentially Exposed Workers[a]	Selected Toxic Exposures[b]	Known or Suspected Health Effects[c]
		Oil mists (from machine tools)	Possible carcinogen, dermatitis
		Solvents, acids, chromium and nickel fumes (from metal-plating operations)	Skin irritation, headache, nervous system disorder, respiratory problems, carcinogens
11. Municipal sewage treatment, landfill, and incinerator workers	1,700	Hydrogen sulfide, carbon monoxide, methane and other gases (from sewage decomposition)	Eye, nose, and throat irritation, respiratory arrest, nervous system disorders
		Solvents, acids, radioactive wastes (from industrial wastewaters)	Nervous system disorders, possible carcinogens
		Liquid chlorine (from sewage plant disinfection)	Skin and lung irritation and burning, respiratory problems
		Volatile organic compounds, heavy metals (from landfill gases)	Nervous system disorders, possible carcinogens
		Lead, cadmium (from incinerator dust and ash)	Nervous system disorders, kidney disease, and hypertension

TABLE 5.2 (Continued)

	Occupation	Estimated Number of Potentially Exposed Workers[a]	Selected Toxic Exposures[b]	Known or Suspected Health Effects[c]
12.	Printing workers	25,100	Oil mists, cleaning solvents, dyes, and cadmium and lead (from inks)	Possible carcinogens, nervous system disorders
13.	Public transit maintenance workers (bus and subway)	15,000	Carbon monoxide, diesel particulates and other motor vehicle pollutants	Cardiovascular problems, lung diseases, lung cancer
			Asbestos (from brake linings)	Asbestosis, lung cancer, cancer of the chest lining or abdomen, other cancers
			Solvents (from graffiti and other cleaning operations)	Nervous system disorders

Sources: [a] New York Department of Environmental Protection, Bureau of Air Resources, Asbestos Control Program (No. 1); New York State Department of Labor, "Occupational Needs in the 1980's: New York City, 1987–1989" (Nos. 2, 4, 5, 8, 9, 10, 12); Bridge and Tunnel Officers Benevolent Association (No. 3); Port Authority of New York and New Jersey (No. 3); New York State Department of Environmental Conservation, Bureau of Pesticides (No. 6); Uniformed Firefighters Association of Greater New York (No. 7); District Council 37, American State, County and Municipal Employees (No. 11); New York City Department of Sanitation (No. 11); Transport Workers Union (No. 13).

[b] International Labour Office, *Encyclopaedia of Occupational Health and Safety*, 3d rev. ed., Vols. 1, 2 (1982) (Nos. 1, 2, 3, 4, 5, 7, 8, 9, 10, 12); Carl Zenz, *Occupational Medicine: Principles and Practical Applications*, 2d ed. (1988) (Nos. 2, 4, 6, 7, 8, 9, 10); R. Gregory Evans *et al.*, "Cross-sectional and Longitudinal Changes in Pulmonary Function Associated with Automobile Pollution Among Bridge and Tunnel Officers," *American Journal of Industrial Medicine*, Vol. 14, No. 1 (1988) (No. 3); Frank B. Stern *et al.*, "Heart Disease Mortality Among Bridge and Tunnel Officers Exposed to Carbon Monoxide," *American Journal of Epidemiology*, Vol. 128, No. 6 (December 1988) (No. 3); City of New York, Mayor's Office of Operations, Citywide Office of Occupational Safety and Health and District Council 37 Education Fund, *Right-to-Know Handbook for Sewage Treatment Workers* (No. 11); Allen Kraut *et al.*, "Neurotoxic Effects of Solvent Exposure on Sewage Treatment Workers," *Archives of Environmental Health*, Vol. 43, No. 4 (July/August 1988) (No. 11); U.S. Department of Health and Human Services, National Institute for Occupational Safety and Health, *Health Hazard Evaluation Report*, HETA 82-305-1541, Fountain Avenue landfill, Brooklyn, New York (December 1984) (No. 11).

[c] *Id.*

main floor work area and more recently in a hidden basement storeroom in which mercury was reclaimed. OSHA inspectors visited the plant in 1981 and 1983 and found serious violations (i.e., contaminated work surfaces, workers eating in areas exposed to mercury) on both occasions. But the cases were settled for nominal fines and Pymm was granted four extensions—until late 1985—to clean up plant operations.

New York City Health Department officials, uncertain about their jurisdiction, peeked into the case in late 1984. They found more than half of Pymm's 90 employees with urine mercury levels ten times above normal. About 20 employees showed indications of nervous system damage associated with mercury toxicity. A third OSHA inspection in December 1984 produced new charges, but they were eventually settled for under $1,000 in penalties.

It was not until October 1985 that OSHA uncovered the poorly ventilated, mercury-contaminated basement operation. Higher fines and a criminal prosecution by the Brooklyn district attorney followed. But for the four-year period from 1981–85, workers at Pymm were subjected to dangerous levels of mercury as OSHA officials continued to boot the issue around.[54]

As the Pymm case suggests, the federal Occupational Safety and Health Administration and its foot soldiers in the New York regional office have generally not been a dominant force in the struggle against workplace toxins. In fact, OSHA's presence has been shrinking in New York City during the 1980s—the Brooklyn area office, one of three in the city, closed in 1982. And according to a 1987 draft report by the Labor Department's inspector general, the New York region's OSHA health and safety program suffered from both systemic weaknesses in office management and routine violations of internal guidelines. The report echoed what labor veterans have long been saying—that OSHA's effectiveness has been compromised by its nonconfrontational approach, often untimely or inadequate inspections, frequently poor case follow-up, and general failure to impose strict penalties even on willful violators. Following release of the inspector general's report and the Pymm case exposé, Labor Department officials reshuffled top managers in OSHA's New York regional offices.[55]

Absent strong federal stewardship, independent resistance fighters have stepped into the fray. Leading unions have added safety and health specialists to their staffs. Private groups such as the New York Committee on Occupational Safety and Health have begun to offer worker education and training programs. And the Mount Sinai Medical Center has christened an occupational health clinic to assist in the prevention and early detection of work-related illnesses.

What a surprise it would be if government agencies really cracked down on occupational toxins in the 1990s.

Epilogue

"The first requisite of a good citizen in this Republic of ours," said Theodore Roosevelt, an environmentalist of sorts in his own time, "is that he shall be able and willing to pull his weight." Throughout this book, we have laid out an agenda for what our elected and appointed officials must do to clean the air, safeguard the waters, and improve the quality of life. But such problems cannot be solved by government alone. We turn now to some of the things that can begin to transform every New Yorker into a roughrider in the campaign to protect this city's environment.

Don't feel left out if you are not a New York resident. Regardless of where you live, you should find, in this epilogue, useful ideas that can be implemented in your community. If you are in doubt as to what agency to contact for help, a good starting point is your state's department of environmental protection or natural resources, or the closest regional office of the U.S. Environmental Protection Agency. A comprehensive state-by-state listing of government agencies and nonprofit organizations dealing with the environment is the indispensable *Conservation Directory* published annually by the National Wildlife Foundation, 1400 Sixteenth Street, N.W., Washington, D.C. 20036-2266 ([703] 790-4000).

SOLID WASTE

On the garbage frontlines, New Yorkers wield enormous power. More important even than the political clout that can delay proposed incinerator projects is the positive contribution all of us can make by recycling more and reducing the amount of waste we produce in the first place.

Here are the first steps you can take to lessen the need for unwelcome incinerators, save precious landfill space, and help bail the city out of its waste disposal bind. Separate newspapers, magazines, and cardboard boxes for recycling. (Sanitation workers are now making special pickups of these items in many neighborhoods; the program will be implemented citywide over the next few years.) The newspapers should be tied with string; glossy advertising supplements and magazines should be separately tied, since the recycling process for them differs. If all New Yorkers started participating in the city's newspaper, magazine, and cardboard program, you'd be surprised how quickly the recycling ethic would spread.

Over the next few years, the Sanitation Department will also be collecting such other recyclables as glass, metals, and plastic. You should participate in those programs and encourage your neighbors to do so as well. To find out when the new measures will be coming to your community (and to press for their early arrival), you can call the nearest Sanitation Department district office (listed in the back of the telephone directory under "New York City Government Offices") or your local community board, whose numbers may be obtained from your borough president's office.

Don't forget that all soda and beer containers can be returned to grocery stores or redemption centers for their nickel deposits. If you are pressed for time, you can probably find a youngster in your apartment building or on your block who would be happy to take these items off your hands. If you live in Manhattan, you can call the

Collection Network at WE CAN, an innovative grass roots organization, to arrange for pickup of your cans and bottles; WE CAN does not handle individual requests, but does collect from businesses, churches, synagogues, schools, and individual apartment buildings. It provides sanitary storage boxes and free pickup; significantly, its profits help support the homeless. WE CAN is planning to expand its operations into Brooklyn. The group can be reached at (212) 262-2222 during regular business hours.

When you go to the grocery store, you can cut down on the number of plastic or paper bags by carrying your own fabric satchel, like the kind that are often seen in Europe. (Among the stores where you can purchase these satchels is Dean and Deluca, 550 Broadway at Prince Street in Manhattan, [212] 431-1691.) And if you're just buying a handful of groceries, you can save time and skip the extra packing by simply saying, "Thanks, but I don't need a bag."

The purchasing decisions New Yorkers make also can have a dramatic impact on the trash crisis. Most important is to start buying products made from recycled materials. You'll soon start seeing products made with recycled items marked with a special logo to help you identify them. You can already find greeting cards printed on recycled paper and you can make a personal start by purchasing them and also by using recycled stationery. (Among the companies that sell recycled paper products is the Earth Care Paper Company, P.O. Box 3335, Madison, Wisconsin 53704, [608] 256-5522.)

Fancy packaging adds to the cost of an item you're buying, but not to its quality. Try to base your purchasing decisions on the product, not its package. And, when buying consumer items, inquire about the durability of the product. A cheaper model that will fall apart more quickly is no bargain.

You can also take a bite out of the trash problem by cutting back on the use of disposable items. Do you really need disposable razors or throwaway plates and utensils? Would you be able to get by using cloth instead of disposable diapers? If you're one of those who uses those new disposable cameras, cigarette lighters, or flashlights, you've got more opportunities than most to help cut down on excessive trash.

The city's new mandatory recycling law requires local officials to come up with a battery recycling program by 1991. You can lessen the nettlesome problem of battery disposal (they release heavy metals when burned and their toxic constituents can leach out of landfills) by switching to rechargeable batteries. They really work. (You can also call the Environmental Action Coalition, [212] 677-1601, for details on its fledgling battery recycling program.)

Leaves and yard waste can account for more than 15 percent of municipal trash in some months. If you have a backyard with sufficient space, you might give thought to creating your own compost pile. Composting returns nutrients to the soil and removes hard-to-burn items from the waste stream. You can get the simple information on how to do this at your public library, or from *Organic Gardening* magazine (Rodale Press, 33 East Minor Street, Emmaus, Pennsylvania 18098).

If you have no room for your own compost pile, you can separately bag your yard wastes for Sanitation Department pickup. Ask your local Sanitation Department district manager whether your yard wastes are being composted, and press local park officials to allocate space for composting operations in a portion of their green space. Your community board should be able to help set up such a neighborhood composting program.

Finally, don't forget your workplace when it comes to recycling and waste reduction. You can ask your employer to purchase recycled products. (Call the Council on the Environment of New York City, [212] 566–0990, for information about recycling your office's white ledger and computer paper.) You can organize newspaper, bottle, and can drop-off points on the job. And you can help convince your employer that it makes good business sense to take the lead in aggressively pursuing environmental programs like recycling and the use of recycled paper.

These suggestions are only the tip of the iceberg. The opportunities to make a difference on solid waste issues are growing every day.

WATERWAYS AND THE COAST

At the other end of the spectrum are concerns of clean waterways and protected shorefronts. These areas present fewer options for individual citizen action. But even here there are things you can do. Why not enlist in the fight against toxic water pollution? Part of the problem can be traced to improper disposal of household toxins. Never pour used motor oil, old paints, photographic chemicals, pesticides, or other harmful products down the drain or into the sewers. Gasoline stations will often take your used oil. You should consider storing your old paint cans and other toxins. Meanwhile, you can work with your local community board to organize a household toxic waste collection day.

Defending New York's waterways and coastal frontier is a challenge for which collective action and neighborhood muscle flexing are necessary. One way to do this is to join, or step up your participation in, a local community or environmental group. On the water pollution front, in addition to the national groups like NRDC, you might hook up with a neighborhood-based organization that is looking after your favorite coastal spot—the Alley Pond Environmental Center ([718] 229-4000), Bronx River Restoration ([212] 933-4079), Coalition for the Bight ([212] 460-9250), Friends of Gateway (National Recreation Area) ([212] 233-4788), and the New York City Audubon Society ([212] 691-7483) are a few of many such groups. You can obtain a comprehensive directory of environmental organizations in New York City from the Natural Resources Group of the city Parks Department ([212] 360-8201).

One smart way to protect yourself and your family and friends until water quality improves is to respect state warnings on the consumption of fish caught in New York waterways. The state Department of Health has set up a toll-free hot line for information on, among other things, which fish from which waters you'd be wise to avoid. The telephone number is (800) 458-1158.

Similar caution should be exercised when it comes to swimming in nondesignated bathing beaches. To safeguard yourself from sewage-carrying bacteria and for more obvious safety reasons, confine your swimming to the roughly 18 miles of official beachfronts in

New York City. Even there, obey posted restrictions on bathing, especially for two days following sewage-releasing heavy rainfalls. For information on bathing conditions at city-run beaches, you can call the city Parks Department's hot line, (718) THE-DEEP.

Some of the most fiery conflicts are now shaping up over coastal development issues. And land use provisions in the new city charter offer opportunities for citizen involvement; the Municipal Arts Society and NRDC recently have published a guide to citizen participation; you can get one from the Municipal Arts Society at (212) 980-1297.

If global warming proceeds unchecked, the first places it will be felt are along the city's coast. There are many things you can do to help slow the greenhouse effect. Number one is energy conservation. It goes without saying that you aid the environment by riding public transit and by choosing autos that get high gas mileage. You can also make sure your home or apartment is well insulated. And, when shopping, purchase energy-efficient appliances; refrigerators and air conditioners are two items with which you can realize big savings. Carefully check the "total dollar operating costs" label on refrigerators; the lower, the better. For room air conditioners, look for the "energy efficiency ratio" (EER) label, and select models with EERs of 10 or higher.

You couldn't find a more beautiful way to fight global warming than to plant and care for urban street trees. (Trees mitigate the problems of global temperature rise by converting carbon dioxide, the main greenhouse gas, into oxygen.) By themselves, the trees New Yorkers plant and care for will have only a tiny impact on global warming. But protecting street trees is a trend that could spread worldwide. Besides, it's amazing how a row of street trees can transform the whole feeling on a block in New York City.

There are two ways to brighten your street with a new tree. If you want the city to do the work, you should present a request to your local community board, but be prepared to wait several years. If you are willing to take on the costs yourself, or as part of your block association, you'll need a permit from the city Parks Department's urban foresters. (Their borough office telephone numbers are: Bronx, [212] 430-1877; Brooklyn, [718] 768-0224; Manhattan, [212] 860-1844; Queens, [718] 520-5315; and Staten Island,

[718] 816-9193.) Information about which trees are best suited for your neighborhood, how to purchase them, and pruning and maintenance advice (as well as facts about city trees that you can find nowhere else) is available from the nonprofit New York City Street Tree Consortium, 16 West 61st Street, New York, New York 10023, (212) 830-7992.

AIR POLLUTION

Even as government officials kick around air pollution issues, there are many things you can do to reduce your own exposures.

Start first with indoor air. We assume that you understand the risks of smoking. If you smoke, you can do nothing better for yourself and your family than to contact your local lung association for information on how to quit. It can't hurt just to talk to these health specialists, so why not call today? (New York Lung Association, which covers Manhattan, Staten Island, and the Bronx, [212] 889-3370; American Lung Association of Brooklyn, [718] 624-8531; Queensboro Lung Association, [718] 263-5656.)

Until you quit, you can do your housemates and colleagues a big favor by keeping your smoking out of the home and out of the office. Nonsmokers, meanwhile, can insist on their new legal rights to a smoke-free work environment. A gentle request to a smoker is often usually all that is necessary in an office (or restaurant, elevator, or movie theater, for that matter). But if problems persist, you should write to the New York City Health Department (125 Worth Street, New York, New York 10013), which enforces the local smoking law. City Council president Andrew Stein has expanded ombudsman powers under the new city charter, and has shown an interest in health issues. You might forward a copy of your complaint to his office (Municipal Building, 15th Floor North, New York, New York 10007).

Here are some other ways to protect the air in your home or apartment. If you live in a private house, consider testing for radon. Relatively inexpensive test kits are available, but check with the state Health Department's toll-free health line ([800] 458-1158) for advice on recommended test products. Be wary of unvented kerosene and gas heaters; if you have a fireplace, be sure chimneys and flues are properly maintained. Use paint strippers and other chemical solvents only in areas that are well ventilated. Forget about air "fresheners" and open a window. Be careful with mothballs (if possible, put them in trunks that can be stored in the attic or garage) and pass over pressed-wood products containing formaldehyde. When cooking, always use exhaust fans over gas stoves.

There are plenty of helpful publications about indoor pollution. Two to start with are: *The Inside Story: A Guide to Indoor Air Quality,* from the U.S. Environmental Protection Agency, Washington, D.C. 20460; and *Radon: A Homeowner's Guide to Detection and Control,* from Consumers Union (publishers of *Consumer Reports*), 256 Washington Street, Mt. Vernon, New York 10553, (914) 667-9400.

Stepping outside, keep your eyes peeled for black smoke pouring out of rooftop stacks. Whether they are from boilers or incinerators, such emissions should not be visible for more than two minutes an hour. If you notice a building with a continuing violation, write to the New York City Department of Environmental Protection, Bureau of Air Resources, documenting your complaint (these offices are scheduled to move to Queens in 1990; telephone ahead to check the correct address before writing). Send a copy of your letter to NRDC.

You can do your part in motor vehicle pollution control by keeping your car well tuned. As a pedestrian, try to avoid inhaling heavy plumes of exhaust. Without getting too carried away about this, you can hold your breath for a moment when caught behind an accelerating bus or truck, and, for long trips, you might want to walk on the side of the street across from the bus lane. It's also best to avoid exercises such as jogging or biking in heavy traffic, especially on hazy summer days. While there are still many obstacles to bicycle commuting in New York City, the nonpolluting bicycle remains (along with walking) the most environmentally sound transportation mode. May the number of law-abiding bike commuters continue to grow. For more information, and an interesting city biking newsletter, contact Transportation Alternatives, 494 Broadway, New York, New York 10012, (212) 941-4600.

Finally, here are three things you can do to protect the ozone layer, the fragile shield about 18 miles above the earth's surface, which protects the planet from the sun's dangerous ultraviolet radiation. First, since leaky automobile air conditioners are the largest source of ozone-destroying chlorofluorocarbons (CFCs), when your air conditioner breaks, don't just add CFCs, get the leak fixed. (And don't go for unnecessary preventive service for your auto air conditioner; that, too, will release CFCs unless your repair shop has CFC-recycling equipment.) Second, skip halon-containing fire extinguishers when purchasing such items for your home; their ozone-depleting contents will eventually leak even if the extinguishers are not used. Third, some aerosol cans still use CFCs and should be avoided. Among them some aerosol camera lens dust removers, and some boat horns. Check labels carefully and pass on products containing CFCs, including:

CFC-11	trichlorofluoromethane
CFC-12	dichlorodifluoromethane
CFC-113	trichlorotrifluoroethane
CFC-114	dichlorotetrafluoroethane
CFC-115	(mono)chloropentafluoroethane
Halon-1211	bromochlorodifluoroethane
Halon-1301	bromotrifluoroethane
Halon-2402	dibromotetrafluoroethane.

DRINKING WATER

Protecting the quality of New York City's drinking water today and conserving water for tomorrow—these two challenges for city water officials are both areas where citizen actions can make a difference.

On the whole, New York City's water supply today is still one of noteworthy quality. But the city's drinking water quality assurance program effectively ends at the distribution mains, which deliver water to service connectors that feed individual homes and apartments. Citizens must take over as water quality watchdogs where the city leaves off.

If you notice persistent taste or odor problems with your tapwater, notify the New York City Department of Environmental Protection in writing and request a water quality test. If the city does not explain the problem to your satisfaction, or if you have other reasons to suspect your drinking water purity, the next step is an independent test of your water by a qualified laboratory. Consumers Union has recommended three firms that can perform such tests by mail: WaterTest, 33 South Commercial Street, Manchester, New Hampshire 03101, (800) 426-8378; National Testing Laboratories, 6151 Wilson Mills Road, Cleveland, Ohio 44143, (800) 458-3330; and Suburban Water Testing Laboratories, 4600 Kutztown Road, Temple, Pennsylvania 19560, (800) 433-6595. Testing costs are often under $100, depending on the type and number of substances under investigation. If a test of your tapwater reveals high levels of a suspected chemical, you might want to conduct a second analysis (with another lab) before purchasing expensive treatment equipment.

Finally, a tip from the better-to-be-safe-than-sorry department. Perhaps 10 percent or more of New York City households may have elevated levels of lead in their tapwater (usually from leaded pipes and lead-soldered joints in home plumbing, not city reservoirs). Lead levels are highest first thing in the morning, after the city's slightly acidic water has sat in household water pipes overnight. Do not drink this so-called first-draw water, or use it to make coffee, tea, juice, baby formula or hot cereal. Rather than letting the water run for several minutes (which can bring down lead levels, but is wasteful), you can simply fill a bottle of water after washing your dinner dishes and keep it in the refrigerator for use with breakfast. By not drinking tapwater that has collected in the household pipes for many hours, you can significantly cut your family's potential exposure to lead from drinking water. And water tasters report that chilled water also tastes better.

National figures suggest that every New Yorker uses about 80 gallons a day of water just in the home. That's at least four to six bathtubs full. There are many ways to conserve this precious liquid. High on the list is to fix leaky faucets and flushers in your home, or to report such problems to your superintendent if you live in an apartment house. Repairing leaking water fixtures will hardly make you feel like the Rachel Carson of environmental protection, but its importance to the city's drinking water future cannot be overemphasized.

Teach your children the commonsense rules—don't leave water running, don't waste water in the bath or shower, and don't ask for a glass of water if you aren't going to drink it. You can also set a good example by installing low-flow shower heads and by putting a plastic container (filled with water and with its screw-on top affixed) into your toilet tank. (This latter measure can save as much as one-half gallon per flush and is a painless way to stretch the city's drinking water supply. In the words of a popular advertising slogan, "just do it.")

TOXICS

Maybe one of our suggestions for protecting New Yorkers from toxic exposures will save a life someday. On the asbestos front, the warnings are simple—carefully check your home and workplace for asbestos. It is most likely to be found in the home as insulation around pipes, boilers, and heating ducts. One helpful pamphlet, *Asbestos in New York City Homes,* can be obtained from the Council on the Environment of New York City, 51 Chambers Street, New York, New York 10007 ([212] 566-0990). Expert advice on asbestos identification and abatement options can be obtained by contacting the Environmental Protection Agency's regional asbestos coordinator, U.S. EPA, Woodbridge Avenue, Edison, New Jersey 08837 ([201] 321-6668).

If you do find asbestos in your home or place of business, don't panic. Probably the worst thing you can do is hire an unskilled laborer to remove it, or to rip it out yourself. Such actions are sure to make matters worse. Sometimes, covering, or encapsulation, of in-place asbestos may be preferable to removal. If the asbestos you have discovered is crumbling or otherwise damaged, call in a state-certified asbestos abatement worker. You can get a listing from the New York State Department of Labor ([718] 797-7659).

There's no excuse for lead poisoning in New York City in the 1990s, and there is much you can do to prevent its occurrence. We have already mentioned a method for reducing risks from lead in drinking water. But the primary exposure route today is leaded paint. If you live in an older home with peeling or flaking paint and suspect that it may be leaded (many household paints contained lead until the late 1970s), childhood blood-lead testing is a must. You can call the city's Health Department, Bureau of Lead Poisoning Control ([212] 334-2611) or local clinics like the well-regarded program at Montefiore Hospital in the Bronx ([212] 920-5016) for information on testing.

City law requires that landlords strip or cover lead paint in all homes or apartments where children under six are living. Another law requires expedited action by landlords whenever a resident child has been diagnosed as having elevated lead levels. If you are aware of a location where such conditions exist, you should write to the city Health Department and insist that its staff seek prompt action. Send copies of your request to City Council president Andrew Stein and to NRDC.

Finally, although lead in gasoline is largely a thing of the past, this toxic metal has accumulated for years in soil near highways and other heavily traveled streets. If your front yard or neighborhood playground is located right on a busy thoroughfare, and you especially have reason to suspect any other source of lead exposure for your child, you might want to arrange for a test of lead levels in the soil. If you are concerned with a public area like a schoolyard or playground, your local community board should be able to assist in getting the Health Department to arrange for the soil sampling.

We leave you with a few thoughts on what you can do to protect yourself from toxic substance exposures on the job. First, learn what chemicals are used at your workplace. Under state and federal law, most employers must provide such information to their employees. The U.S. Occupational Safety and Health Administration is responsible for enforcing the law covering employees in private industry. If your workplace is located in Manhattan, Brooklyn, or Staten Island, contact OSHA's office at 90 Church Street in Manhattan ([212] 264-9840); if you work in Queens or the Bronx get in touch with the office at 4240 Bell Boulevard, Bayside, New York 11361 ([718] 279-9060). The New York State Labor Department investigates nondisclosure complaints filed by city, state, or federal employees. The department's New York City office is at 1 Main Street, Brooklyn, New York 11201 ([718] 797-7669 or 7671). If you encounter problems, contact your union representatives, many of whom have hired environmental health specialists in recent years. District Council 37 (American Federation of State, County and Municipal Employees, AFL-CIO), which represents many municipal employees, has an especially active Safety and Health Unit ([212] 815-1700).

Your family physician may not always be able to provide expert diagnosis and treatment of occupational diseases. If you suspect that you may have a job-related illness, you should consider consulting the nationally respected experts at the Division of Environmental and Occupational Medicine at the Mount Sinai Medical Center ([212] 241-6173).

Notes

Part One: Solid Waste

1. **19,000 tons:** This figure is based upon data from the following sources: personal communication with Beth Hurley, Bureau of Waste Disposal, New York City Department of Sanitation (December 4, 1989); Ivan Braun, Deputy Director, Recycling Programs and Planning Division, New York City Department of Sanitation (December 7, 1989); New York City Department of Sanitation, *Final Environmental Impact Statement for the Proposed Resource Recovery Facility at the Brooklyn Navy Yard* (June 1985), at 1-4; **five pounds:** New York State Department of Environmental Conservation, *New York State Solid Waste Management Plan 1987–1988 Update* (March 1988), at 1-8; **typical Japanese etc.:** Joanna D. Underwood, Allen Hershkowitz, Maarten de Kadt, *Garbage: Practices, Problems and Remedies* (1988), at 3.
2. New York City Department of Sanitation, Bureau of Waste Disposal, "Loads and Tonnage Report" (Summary Report December 1988), at 3, and (October 1989) at 2.
3. **Unsightly swamps:** See, for example, Robert Moses, *Public Works: A Dangerous Trade* (New York: McGraw-Hill, 1970), at 31; **city-operated landfills:** New York City Department of Sanitation, *The Waste Disposal Problem in New York City: A Proposal for Action*, Vol. 1 (April 1984), at I-1.
4. New York City Department of Sanitation, *Fresh Kills Landfill Preliminary Draft Environmental Impact Statement* (December 1985), at ES-8; New York City Department of Sanitation, *Landfill Design Report, Edgemere Landfill*

(November 1985), at 2-19; New York City Department of Sanitation, *Landfill Design Report, Fountain Avenue Landfill* (November 1985), at 2-15; New York City Department of Sanitation, *Landfill Design Report, Pennsylvania Avenue Landfill* (November 1985), at 2-12; City of New York, *Study of Six Department of Sanitation Landfill Disposal Facilities and Their Compliance with the Applicable Federal and State Laws, Rules and Requirements* (January 1980), Appendix D, at 15.

5. **Four city landfills (Pelham Bay, Edgemere, Pennsylvania Avenue, and Fountain Avenue):** Consulting firm of Parsons Brinckerhoff-Cosulich, *Investigation of Indicator Pollutant Levels at New York City Landfills* (June 2, 1982), at 23, 24, 35, 42; **lead, nickel, and PCBs:** Robert C. Ahlert *et al.*, *Evaluation of Landfill Closure Requirements*, U.S. Interior Department, National Park Service, Gateway National Recreation Area (November 1985), at 37-41; **Jamaica Bay:** see, generally, U.S. Department of Interior, National Park Service, Gateway National Recreation Area, Office of Resource Management and Compliance, *Summary of Several Investigations Regarding the Natural Resources of Jamaica Bay and Their Relationship to Environmental Contaminants* (March 1987).

6. New York City Department of Sanitation, *Landfill Design Report, Pennsylvania Avenue Landfill Report* (November 1985), at 1-12–1-16; New York City Department of Sanitation, *Landfill Design Report, Fountain Avenue Landfill* (November 1985), at 1-14–1-18; New York City Department of Sanitation, *Landfill Design Report, Edgemere Landfill* (November 1985), at 1-21–1-27; also Robert C. Ahlert *et al.*, *Evaluation of Landfill Closure Requirements*, U.S. Interior Department, National Park Service, Gateway National Recreation Area (November 1985), at 29-37.

7. **Brookfield Avenue landfill:** New York City Department of Health, *Report on the Brookfield Health Survey* (1983), at 13, 19-20; **Fountain Avenue landfill:** National Institute for Occupational Safety and Health, "Health Hazard Evaluation Report, Fountain Avenue Landfill, Brooklyn, New York" (December 1984), at 1.

8. N.Y. Envtl. Conserv. Law §27-0703 (McKinney 1984); 6 N.Y.C.R.R. Part 360.8.

9. 42 U.S.C. §6949(c). EPA proposed these landfill rules in August 1988. 53 Fed. Reg. 33314 (August 30, 1988).

10. New York State Department of Environmental Conservation, *Inactive Hazardous Waste Disposal Sites in New York State*, Vol. 2 (December 1987).

11. U.S. Environmental Protection Agency, Region II, "Compliance Monitoring Report, Fresh Kills Landfill, Staten Island, New York" (December 17, 1987).

12. **1985 agreement:** *In the Matter of Alleged Violations of Environmental Conservation Law.* Sections 27-0707, 17-0501, and 25-0401, and 6 N.Y.C.R.R. Parts 360, 751, and 661 by the City of New York, DEC File No. 2-0527 (Order on Consent, signed December 16, 1985); **federal court orders:** see, for example, *Township of Woodbridge v. City of New York*, Civil Action No. 79-1060 (D.N.J. Consent Order entered December 7, 1987).

13. **1985 projection:** New York City Department of Sanitation, *Final Environmental Impact Statement for the Proposed Resource Recovery Facility at the Brooklyn Navy Yard* (June 1985), at 1-5; **recent estimates:** New York City

Department of Sanitation, "Fresh Kills Landfill Depletion (FY1990–FY2029)," materials presented to the City Council (March 1989).

14. New York State Department of Environmental Conservation, *Inactive Hazardous Waste Disposal Sites in New York State*, Vol. 2 (December 1987), at 2-21.

15. **1991 phase out:** *In the Matter of the Development and Implementation of a Remedial Investigation, Feasibility Study and Implementation of a Interim and Final Remedial Program for an Inactive Hazardous Waste Disposal Site Under Article 27, Title 13 of the Environmental Conservation Law of the State of New York and of the Interim Operation of a Solid Waste Management Facility Under Article 27, Title 7, by the City of New York*, DEC File No. D2–7001–87–07 (Order on Consent, signed August 19, 1987), at 6; **ash disposal:** see New York City Department of Sanitation, *A Waste Disposal Problem in New York City: A Proposal for Action*, Vol. 1 (April 1984), at 6; see also Bureau of National Affairs, *Environment Reporter: Current Developments*, "New York City Agrees With State to Close Landfill" (July 17, 1987), at 838.

16. **Hazardous waste list:** New York State Department of Environmental Conservation, *Inactive Hazardous Waste Disposal Sites in New York State*, Vol. 2 (December 1987), at 2-7; **PCBs:** U.S. Interior Department, National Park Service, Gateway National Recreation Area, Office of Resource Management and Compliance, *Summary of Several Investigations Regarding the Natural Resources of Jamaica Bay and Their Relationship to Environmental Contaminants* (March 1987), at 1; **cleanup agreement:** *In the Matter of Alleged Violations of Environmental Conservation Law.* Sections 27-0707, 17-0501, and 25-0401 and 6 N.Y.C.R.R. Parts 360, 751, and 661 by the City of New York, DEC File No. 2-0954 (Order on Consent, signed December 16, 1985), at 8.

17. **Waste oil:** Robert C. Ahlert *et al.*, *Evaluation of Landfill Closure Requirements*, U.S. Interior Department, National Park Service, Gateway National Recreation Area (November 1985), at 48-49; **oozing into Jamaica Bay:** *Id.* at 3-4; see also New York State Department of Environmental Conservation, *Inactive Hazardous Waste Disposal Sites in New York State*, vol. 2 (December 1987), at 2-5.

18. **Waste oil deposits:** Robert C. Ahlert *et al.*, *Evaluation of Landfill Closure Requirements*, U.S. Interior Department, National Park Service, Gateway National Recreation Area (November 1985), at ix, 49-51; **consent agreement:** *In the Matter of Alleged Violations of the Environmental Conservation Law.* Sections 27-0707, 17-1501, and 25-0401, and 6 N.Y.C.R.R. Parts 360, 751, and 661: by the City of New York, DEC File No. 2-0953 (Order on Consent, signed December 16, 1985), at 3. A second city/state consent agreement requires the Department of Sanitation to remove or safely contain waste oils and other hazardous materials illegally dumped at the site. No final deadline for city action, however, is contained in the agreement. *In the Matter of Alleged Violations of the Environmental Conservation Law.* Sections 27-0707, 17-0501, and 25-0401, the Navigation Law, Section 173 and 6 N.Y.C.R.R. Parts 751 and 661 by the City of New York, DEC File No. 2-0953A (Order on Consent, signed December 16, 1985).

19. **Millions of gallons of industrial waste:** New York State Department of Envi-

ronmental Conservation, *Inactive Hazardous Waste Disposal Sites in New York State*, Vol. 2 (December 1987), at 2-31; **clay cap:** personal communication with Jim Meyer, Deputy Director of Public Policy, Office of Resource Recovery, New York City Department of Sanitation (September 13, 1988); **state environmental rules:** *In the Matter of Alleged Violations of Environmental Conservation Law.* Sections 27-0707, 17-0501, and 25-0401, and 6 N.Y.C.R.R. Parts 360, 751, and 661 by the City of New York, DEC File No. 2-0952 (Order on Consent, signed December 4, 1985).

20. **Illegal burials and state listing:** New York State Department of Environmental Conservation, *Inactive Hazardous Waste Disposal Sites in New York State*, Vol. 2 (December 1987), at 2-1; **Health Department investigation:** New York City Department of Health, Division of Community and Occupational Health Promotion, "An Evaluation of Childhood Leukemia in the Pelham Bay Area of the Bronx" (October 1988), at 2; **consent order:** *In the Matter of Alleged Violations of Environmental Conservation Law.* Sections 27-0707, 17-0501, and 25-0401, and 6 N.Y.C.R.R. Parts 360, 751, and 661 by the City of New York, DEC File No. 2-0956 (Order on Consent, signed December 4, 1985).

21. **Lead in soil:** New York City Department of Health, Division of Environmental Toxicology, "Public Health Impact of Exposure to Soil at Fairfield Estates" (February 1987), at 5, 16; **Idlewild landfill:** New York City Public Development Corporation, "Report on Investigation of Possible Contamination of the Idlewild Park-Air Cargo Terminal Site" (September 1986), at 14-16, 32-33; **rercommended safety measures:** New York City Department of Health, Division of Environmental Toxicology, "Public Health Impact of Exposure to Soil at Fairfield Estates" (February 1987), at 22; see also New York City Public Development Corporation, "Draft Environmental Impact Statement, Air Cargo Industrial Park, Springfield Gardens-Brookville, Queens, New York" (May 1986).

22. **4,200 pieces:** Personal communication, Arthur Ashendorff, Director, Bureau of Public Health Engineering, New York City Department of Health (March 8, 1989); **1 to 10 percent and beach closings:** New York State Department of Environmental Conservation, *Investigation: Sources Of Beach Washups in 1988* (December 1988).

23. **Over 20 tons a day:** New York State Department of Health, *Infectious Waste: A Statewide Plan for Treatment and Disposal (Preliminary)* (August 1988), Appendix II, at 26 (This number is based on a 127-ton weekly incineration rate, averaged over 302-day calendar year.); **new or proposed incinerators:** City of New York, Department of Environmental Protection, Bureau of Air Resources, "Installation Permit Status of New York City Medical Care Waste Incinerators" (February 23, 1989).

24. **Four times higher plastics:** R. J. Allen *et al.*, "Air Pollution Emissions from the Incineration of Hospital Waste," *Journal of the Air Pollution Control Association*, Vol. 36, No. 7 (July 1986), at 829; see also Jack D. Lauber, "An Overview of the Burning Issues of Municipal and Hospital Waste Incineration," at 6, presented at the University of Florida, Gainesville, Florida, March 1, 1986, at "There Is No Away," Interdisciplinary Conference On Strategies for the Comprehensive Management of Domestic, Municipal and

Hazardous Wastes; **dioxin:** David Marrack, "Hospital Red Bag Waste: An Assessment and Management Recommendations," *Journal of the Air Pollution Control Association*, Vol. 38, No. 10 (October 1988), at 1,310; see also Jack D. Lauber, "New Perspectives on Toxic Emissions from Hospital Incinerators," Solid Waste Management Conference on Solid Waste Management and Materials Policy, New York, New York, February 12, 1987.

25. **Bottle bill:** N.Y. Envtl. Conserv. Law §§27-1005 *et seq.* (McKinney 1984); **compromise bill:** Solid Waste Management Act of 1988 (Chapter 70, Laws of 1988).

26. Administrative Code of the City of New York §§16-301 *et seq.*

27. B. Commoner *et al.*, *Development and Pilot Test of an Intensive Municipal Solid Waste Recycling System for the Town of East Hampton* (Final Draft), at i-iii.

28. Little information is available on the extent of recycling efforts by private firms that are not connected to city or city-sponsored recycling programs. These operations remove refuse (i.e., scrap metal, waste paper, or construction debris) before it enters the city's waste stream. The city has variously estimated, based on regional and national studies, that between 1,000 and 4,000 tons per day or more of such private sector recycling may be taking place. City officials have, however, acknowledged that there is considerable uncertainty associated with these numbers. See, for example, City of New York, Department of Sanitation, *A Status Report on Materials Recycling Activities in New York City* (December 1985), at Appendices I-A, I-D. See also City of New York, Department of Sanitation, *The Waste Disposal Problem in New York City: A Proposal for Action*, Vol. 7 (April 1984), at II-8.

29. **450 tons:** Personal communication with Ivan Braun, Deputy Director, Recycling Programs and Planning Division, New York City Department of Sanitation (December 7, 1989); **litter and landfill space:** Nelson A. Rockefeller Institute of Government, *The New York Returnable Beverage Container Law: The First Year* (March 15, 1985), at 145-46, 176-80.

30. Franklin Associates, *The Fate of Used Beverage Containers in the State of New York, Summary* (July 1986), at S-7, S-16.

31. **Aluminum cans and glass bottles:** Nelson A. Rockefeller Institute of Government, *The New York Returnable Beverage Container Law: The First Year* (March 15, 1985), at 180-90; see also New York State Department of Environmental Conservation, *New York State Solid Waste Management Plan 1987–1988 Update* (March 1988), at 3-4; **plastic bottles:** Cynthia Pollock, "Mining Urban Wastes: The Potential for Recycling," Worldwatch Paper 76 (April 1987), at 27; **statewide recycling rates:** New York State Department of Environmental Conservation, *New York State Solid Waste Management Plan 1987–1988 Update* (March 1988), at 3-4.

32. **Apartment house program:** Personal communication with Ivan Braun, Deputy Director, Recycling Programs and Planning Division, New York City Department of Sanitation (December 7, 1989); **apartment house programs:** personal communication with Stephen Gallagher, Waste Management Project Director, Environmental Action Coalition (December 6, 1989); and Ivan Braun, Deputy Director, Recycling Programs and Planning Division, New York City Department of Sanitation (December 7, 1989).

33. **1987 City Council action:** Administrative Code of the City of New York §6-122; **handful of contracts:** New York City Department of General Services, "Report on Recycled Paper Procurement Activities" (April 29, 1989); **less than enthralled:** see, e.g., testimony of Hadley Gold, Commissioner, New York City Department of General Services, before New York City Council Environmental Protection Committee, March 13, 1989.

34. **Double the trash:** Joanna D. Underwood, Allen Hershkowitz, and Maarten de Kadt, *Garbage: Practices, Problems and Remedies* (1988), at 3; **25 percent more by 1997:** New York State Department of Environmental Conservation, *New York State Solid Waste Management Plan 1987–1988 Update* (March 1988), at 2-11–2-12, 7-5.

35. On May 24, 1989, the New York State Supreme Court in Suffolk County upheld the statute on constitutional grounds but ruled that the county must complete a full environmental review before the law could take effect. *The Society of Plastics Industry, Inc. et al. v. The County of Suffolk, et al.*, No. 88-11262 (N.Y. Sup. Ct., Suffolk Co., May 24, 1989).

36. **One-third of total garbage:** New York State Department of Environmental Conservation, *New York State Solid Waste Management Plan, 1982–88 Update* (March 1988), at 2-11; **fast-growing plastics and 30 percent by volume:** Jeanne Wirka, *Wrapped in Plastics: The Environmental Case for Reducing Plastics Packaging*, Environment Action Foundation Report (August 1988), at 9, 40; **technological hurdles:** Cynthia Pollock, "Mining Urban Wastes: The Potential for Recycling," Worldwatch Paper 76 (April 1987), at 11; **lower return rate:** New York State Department of Environmental Conservation, *New York State Solid Waste Management Plan 1987–1988 Update* (March 1988), at 3-4.

37. Marjorie J. Clarke, Program Director, Municipal Solid Waste Research, INFORM, "Improving Environmental Performance of MSW Incinerators," presented at the Industrial Gas Cleaning Institute Conference (Washington, D.C., November 3–4, 1988), at 10, 42-43.

38. **First refuse incinerator:** City of New York, Department of Sanitation, *Final Environmental Impact Statement for the Proposed Resource Recovery Facility at the Brooklyn Navy Yard* (June 1988), at 1-2, 1-3; **Supreme Court prohibition:** *New Jersey v. City of New York*, 283 U.S. 473 (1931); **17,000 apartment house incinerators and 22 municipal plants:** City of New York, Department of Sanitation, *Final Environmental Impact Statement for the Proposed Resource Recovery Facility at the Brooklyn Navy Yard* (June 1985), at 1-3, 3-2, and "Response to Comments," at 94.

39. New York City Department of Sanitation, *Final Environmental Impact Statement for the Proposed Resource Recovery Facility at the Brooklyn Navy Yard* (June 1985), at 1-4; New York City Department of Environmental Protection, "On-Site Apartment House Incineration in New York City" (1987), at 3, 4.

40. **500-foot stack:** City of New York, Department of Sanitation, *Final Environmental Impact Statement for the Proposed Resource Recovery Facility at the Brooklyn Navy Yard* (June 1985), at 1-1; **935 tons of ash:** *Id.* at 2-101.

41. **Infinitesimal particles:** Calvin R. Brunner, *Hazardous Air Emissions from Incineration* (New York: Chapman and Hall, 1985), at 38; **lodge permanently in the lung:** Frederica P. Perera and A. Karim Ahmed, *Respirable Particles:*

Impact of Airborne Fine Particulates on Health and the Environment (Cambridge, Mass.: Ballinger Publishing Co., 1979), at 31-38; **metals carried on particles:** Marjorie J. Clarke, "Improving Environmental Performance of MSW Incinerators" (INFORM, November 1988), at 9; **lead:** see, *e.g.*, Herbert Needleman *et al.*, "Deficits in Psychologic and Classroom Performances of Children With Elevated Dentine Lead Levels," *New England Journal of Med.* Vol. 300 at 689 (1979); see also James L. Pirkle, Joel Schwartz, J. Richard Landis, and William R. Harlan, "The Relationship Between Blood Lead Levels and Blood Pressure and Its Cardiovascular Risk Implications," *American Journal of Epidemiology*, Vol. 121, No. 2 (1985), at 246; **mercury:** see, *e.g.*, John Doull, Curtis D. Klaassen, and Mary O. Amdur, eds., *Casarett and Doull's Toxicology: The Basic Science of Poisons*, 2d ed. (New York: Macmillan Publishing Co., 1980), at 421-28.

42. **Most toxic synthetic chemical:** Brunner, *supra* at 57; **potent cancer-causing agent:** Stephen C. Schwartz and Peter L. Wolfe, eds., *Dioxin and Resource Recovery* (New York: American Society of Civil Engineers, February 1987), at 19, 21.

43. **Reports from elsewhere:** Seymour Calvert and Harold M. Englund, *Handbook of Air Pollution Technology* (New York: John Wiley & Sons, 1984), at 528; **more toxic pollutants than hazardous waste burners:** Jack D. Lauber, "Toxic Emissions from Small Incinerators" (presented at New York Academy of Sciences, November 19, 1986), at 9-10; **maximum cancer risks:** U.S. Environmental Protection Agency, *Municipal Waste Combustion Study: Report to Congress* (June 1987), at 88.

44. U.S. Environmental Protection Agency, *Municipal Waste Combustion Study: Report to Congress* (June 1987), at 65, 69, 83.

45. Lauber, *supra* note 43 at 5-6; see also Richard A. Denison and Ellen K. Silbergeld, "Risks of Municipal Solid Waste Incineration: An Environmental Perspective," in *Risk Analysis*, Vol. 8, No. 3 (September 1988), at 343, 351-52.

46. See, e.g., Environment Canada, *National Incinerator Testing and Evaluation Program: Environmental Characterization of Mass Burning Incinerator Technology at Quebec City, Summary Report* (June 1988), at 41.

47. City of New York, Department of Sanitation, *Final Environmental Impact Statement for the Proposed Resource Recovery Facility at the Brooklyn Navy Yard*, Appendix D at 2-101.

48. **Properties of toxic waste:** Environmental Defense Fund, "Summary of EP Leaching Tests Performed on Ash from MSW Incinerators" (March 1987); **ash from six incinerators:** New York State Department of Environmental Conservation, Division of Solid and Hazardous Waste, *Ash Residue Characterization Project, Summary Report* (July 1987), at 3, Table 2; **dangerous levels of dioxins:** Richard A. Denison, Environmental Defense Fund, "Fundamental Objectives of Municipal Solid Waste Incinerator Ash Management" (prepared for presentation at the 81st Annual Meeting and Exhibition of the Air Pollution Control Association, June 20-24, 1988), at 2.

49. See, generally, 6 N.Y.C.R.R. Part 360.

50. **New incinerator rules:** 52 Fed. Reg. 25,399 (July 7, 1987); **legislative proposals:** see, e.g., H.R. 4902, 106th Cong., 2d Sess. (1988); S.1894, 100th Cong., 2d Sess. (1988).

Part Two: Waterways and the Coast

1. For a good summary of the history of water pollution control, see Martin Lang *et al.*, "Control of Water Pollution in New York City," *Municipal Engineers Journal*, Vol. 60 (1974), at 117-27.

   Before the construction of the Coney Island plant, the city had relied for several decades only on rudimentary screening devices to remove large debris from a portion of the city's sewage flows. City of New York, Department of Environmental Protection, *Industrial Pretreatment Program, Task 4, Technical Information Review* (May 1983), at 10.

2. City of New York, Department of Environmental Protection, Bureau of Wastewater Treatment, *New York Harbor Water Quality Survey: 1986* (November 1987), at D-1, D-8, D-9, and appendixes D and E.

3. **Permanently banned:** The East River, upper East River, Harlem River, Hudson River, upper New York Bay, the Kill Van Kull, Arthur Kill, and Jamaica Bay, including Plum, Howard, and Canarsie beaches and all of Broad Channel. The City of New York, Office of the Mayor, Press Release (June 23, 1988); **areas not recommended:** Bronx: from Locust Point (East 177th Street) to Throg's Neck Point; Brooklyn: Seagate; Queens: Little Neck Bay, from Fort Totten to Nassau County line; and Staten Island: from Fort Wadsworth to Slatter Avenue, including South and Graham beaches; from Holton Blvd. to Manhattan Street (but not including any areas within 500 feet of polluting outfalls, which are designated as restricted areas, at Joline Avenue, Bedell Avenue, and Lemon Creek/Bayview Avenue). *Id.;* **water bacteria counts:** New York City Department of Health, *Beach and Harbor Water Sampling Program—1986*, at 11.

4. **Closures of oyster beds:** U.S. Environmental Protection Agency, *Stressed Water Evaluations for the Hudson-Raritan Estuary* (February 1984), at 111-14; David R. Franz and William H. Harris, *Final Report: Benthos Study, Jamaica Bay Wildlife Refuge, Gateway National Recreation Area, Brooklyn, New York* (1985), at 2-3; **coliform bacteria concentrations:** 6 N.Y.C.R.R. §41.1; see also Interstate Sanitation Commission, *1988 Annual Report*, at 2, 49-50; **still harvest:** see, for example, "Illegal Taking of Clams from Tainted Bay Rises," *New York Times*, September 21, 1986.

5. **Consumes oxygen:** U.S. Congress, Office of Technology Assessment, *Wastes in Marine Environments* (April 1987), at 90-91; **occasional fish kills:** *Id.* at 22-23; and see, for example, Sam Howe Verhovek, "School of Fish Dies in Bronx Waters," *New York Times*, June 7, 1988, and Philip S. Gutis, "Jersey Fish Kill Stirs New Fears of Decline in Region's Waters," *New York Times*, July 3, 1988; **less serious declines:** New York State Department of Environmental Conservation, Division of Water, Bureau of Monitoring Assessment, *New York State Water Quality 1988* (April 1988), at I-10, III-44–III-51, III-154–III-155.

6. **Primary goals:** 33 U.S.C. §1251(a)(1), (2); **secondary treatment:** 33 U.S.C. §1311(b) (1) (B); secondary treatment regulations are spelled out in 40 C.F.R. 133.012 and N.Y. Envtl. Conserv. Law §17-0509 (McKinney 1984); **classify**

water bodies: 33 U.S.C. §1313; see also N.Y. Envtl. Conserv. Law §§17-0301 (McKinney 1984) and 6 N.Y.C.R.R. §§700-705, establishing New York State's water classification scheme. New York City is also subject to water classifications set by the Interstate Sanitation Commission, a tri-state enviroinmental agency that regulates surface waters in the metropolitan region. N.Y. Envtl. Conserv. Law §21-0501 (McKinney 1984).

7. New York City Department of Environmental Protection, *Monthly Operating Efficiency* (January 1989–June 1989).

8. City of New York, *The Mayor's Management Report* (Preliminary, January 30, 1986), at 139. City officials have also suggested for many years that weak or diluted sewage flows to the plants may contribute to the inability of these plants to achieve secondary standards. See City of New York, Department of Environmental Protection, *Areawide Waste Treatment Management Planning Program*, Section 208 (April 1979), at 2-2.

9. **Reconstruction:** New York City Department of Environmental Protection, *Response to EPA's Comments on a Special Permit Application for the Disposal of Sewage Sludge from Fourteen New York City Water Pollution Control Plants at the Deepwater Municipal Sludge Dump Site* (January 15, 1988), at 12-5–12-6; **lawsuit by New York State's attorney general:** *State of New York v. City of New York*, No. 33707/86 (N.Y. Sup. Ct.) (Judgment on consent) (June 23, 1988); *State of New York v. City of New York*, No. 25296/86 (N.Y. Sup. Ct.) (Judgment on consent) (June 23, 1988); *State of New York v. City of New York*, No. 196/88 (N.Y. Sup. Ct.) (Judgment on consent) (June 23, 1988); **city's own forecast:** City of New York, *The Mayor's Management Report* (Preliminary, January 30, 1987), at 137; **budget changes:** City of New York, *The Mayor's Management Report* (Preliminary, February 15, 1989), at 117; City of New York, Office of Management and Budget, The City of New York Executive Budget Fiscal Year 1990 (May 18, 1989), at 184.

10. New York City, Department of Environmental Protection, *Monthly Operating Efficiency* (January 1989–June 1989).

11. Approximately 70 percent of the city's 6,200 miles of sewage pipes are combined; the remaining 30 percent carry household wastes and stormwater runoff in separate sewer mains. City of New York, Department of Environmental Protection, *Citywide Combined Sewer Overflow Study*, Final Report (November 1986), at 2-2–2-3; **2 to 20 times:** City of New York, Department of Environmental Protection, *Citywide Combined Sewer Overflow Study*, Interim Report (September 1985), at 3-8, Table 5; John J. Roswell, "Combined Sewer Overflow Control in the City of New York," City of New York, Department of Environmental Protection (June 1985; updated September 1987), at 1; **500 open drainpipes:** City of New York, Department of Environmental Protection, *Citywide Combined Sewer Overflow Study*, Final Report (November 1986), at Table 2; see also Interstate Sanitation Commission, *Combined Sewer Outfalls in the Interstate Sanitation District* (October 1988), at xv.

12. **560 million gallons:** City of New York, Department of Environmental Protection, *Citywide Combined Sewer Overflow Study*, Final Report (November 1986), at Table 5; **even after light rains:** *Id.* at Table 6 (these figures are based on an average 6.67-hour rainstorm); **"the major source of contamination":**

City of New York, Department of Environmental Protection, *Citywide Combined Sewer Overflow Study,* Interim Report (September 1985), at 1-1.

13. John J. Roswell, "Combined Sewer Overflow Control in the City of New York," City of New York, Department of Environmental Protection, (June 1985; updated September 1987), at 2-3.

14. **60 million gallons:** New York City Department of Environmental Protection, Bureau of Water Pollution Control, *Flushing Bay Water Quality Facility Plan, Task 5.1 Develop and Evaluate Alternatives, Screening and Preliminary Evaluation* (March 1986), at 2-1C, 2-2A; see also New York City Department of Environmental Protection, *Task 3.2 Report Measurement as Evaluation of Discharge and Pollutant Loadings from Combined Sewer Outflows to Flushing Bay and Creek* (June 1986), at 2-1; **even during dry weather:** New York City Department of Environmental Protection, Bureau of Water Pollution Control, *Flushing Bay Water Quality Facility Plan, Task 5.2 Technical Memorandum Dry Weather Overflow Abatement in Flushing Bay and Creek,* Final Report (November 1987), at 4 (this figure is based on a 24-hour period).

15. **Official monitoring:** See, most recently, New York City Department of Environmental Protection, Bureau of Wastewater Treatment, *New York Harbor Water Quality Survey: 1987* (February 1989), at 6, B-1–B-6; **bacteria counts:** *Id.* at D-2–D-3, E-2; **buildup of odor-producing sewage:** see, for example, Joseph P. Fried, "Split Over Flushing Bay Cleanup Plan," *New York Times,* April 5, 1987.

16. New York State Department of Environmental Conservation, "Interim Procedures, New York Harbor—Wasteload Allocation Analysis" (June 1986), at 1, 4.

17. New York State Department of Environmental Conservation, *PCB Concentrations in the Striped Bass from the Marine District of New York State* (April 1988), at Table 5, Table 25; New York State, Department of Environmental Conservation, *PCB in Hudson River Striped Bass: Ten Years of Monitoring* (June 1988), at Table 3.

18. **Issued warnings:** New York State Department of Health, *1987–1988 Health Advisory* (July 13, 1987); **banned the commercial taking:** 6 N.Y.C.R.R. §§11.1, 11.2, 11.3; **primary nonoccupational source:** U.S. Department of Commerce, National Oceanic and Atmospheric Administration, *Chemical Pollution of the Hudson Raritan Estuary* (December 1984), at 55; Thomas Belton, Robert Roundy, and Neil Weinstein, "Urban Fisherman: Managing the Risks of Toxic Exposure," *Environment,* Vol. 28, No. 9 (November 1986), at 19; **recreational anglers:** see, U.S. Department of Interior, National Park Service, Gateway National Recreation Area, Office of Resource Management and Compliance, *Summary of Several Investigations Regarding the Natural Resources of Jamaica Bay and Their Relationship to Environmental Contaminants,* at 2, Section XII; also, Charles A. Heatwole and Niels C. West, "Shore-Based Fishing in New York City," unpublished paper (November 1984), at 1.

19. **Cadmium, mercury:** Ronald Sloan and Ralph Karcher, "On the Origin of High Cadmium Concentrations in Hudson River Blue Crab (*Callinetes sapidus* Rathbun)," *Northern Environmental Science,* Vol. 3, Nos. 3/4 (1984), at 221-31; Hudson River Foundation, "A Study of the Occurrence of Liver

Cancer in Atlantic Tomcod (*Microgadus tomcod*) from the Hudson River Estuary" (April 1986), at 3-13; **dioxins:** P. O'Keefe *et al.*, "Tetrachloro-dibenzo-*p*-dioxins and Tetrachlorodibenzofurans in Atlantic Coast Striped Bass and in Selected Hudson River Fish, Waterfowl and Sediments," *Chemosphere*, Vol. 13, No. 8 (1984), at 849-60; Thomas J. Belton *et al.*, "A Study of Dioxin (2,3,7,8-Tetrachlorodibenzo-*p*-Dioxin) Contamination in Select Finfish, Crustaceans and Sediments of New Jersey Waterways," New Jersey Department of Environmental Protection, Office of Science and Research (October 30, 1985), at 1-3.

20. **Four toxic metals:** City of New York, Department of Environmental Protection, Bureau of Wastewater Treatment, *New York Harbor Water Quality Survey: 1987* (February 1989), at 11-14, H-1–H-2; **levels of heavy metals:** City of New York, Department of Environmental Protection, Bureau of Wastewater Treatment, *New York Harbor Water Quality Survey: 1986* (November 1987), at Appendix Q; City of New York, Department of Environmental Protection, Bureau of Water Pollution Control, *New York Harbor Water Quality Survey: 1983* (1984), at 1; **organic compounds:** City of New York Department of Environmental Protection, Bureau of Wastewater Treatment, *New York Harbor Water Quality Survey: 1986* (November 1987), at 13-14, and Appendices K, L; New York State Department of Environmental Conservation, *Report of the Fixed Station Toxic Parameter Water Quality Surveillance Network—1986* (January 1988), at 159-70.

21. **DDT, PCBs, and lead:** Robert U. Ayres and Samuel R. Rod, "Patterns of Pollution in the Hudson-Raritan Basin," *Environment*, Vol. 28, No. 4 (May 1986), at 14-43; **toxins linger:** in portions of the Hudson River off New York City, for example, the state has singled out sediments as a primary source of toxic water contamination. New York State Department of Environmental Conservation, Division of Water Bureau of Monitoring and Assessment, *New York State's Water Quality 1988* (April 1988), at III-27, III-140.

22. U.S. Department of Commerce, National Oceanic and Atmospheric Administration, *The National Status and Trends Program for Marine Environmental Quality* (May 1987). See also U.S. Department of Commerce, National Oceanic and Atmospheric Administration, *Report to the Congress on Ocean Pollution, Monitoring, and Research* (August 1989) at 31-32.

23. New York State Department of Environmental Conservation, "Listing of SPDES Permits in New York City" (June 21, 1989).

The three industrial firms are the Amstar Corporation (Brooklyn), Procter & Gamble Manufacturing Company (Staten Island), and AT&T Nassau Metals Corporation (Staten Island). Eight of the utility facilities are Consolidated Edison generating stations: Arthur Kill (Staten Island), Astoria (Queens), East River (Manhattan), 59th Street (Manhattan), Hudson Avenue (Brooklyn), 74th Street (Manhattan), Ravenswood (Manhattan), Waterside (Manhattan). The two other utilities are the Long Island Lighting Company's Far Rockaway Power Station (Queens) and the New York Power Authority's Charles Poletti Power Project (Queens).

24. **800 industrial and commercial firms:** New York City Department of Environmental Protection, *Sludge Management Plan, New York City Industrial Pretreatment Program* (October 1989), at 3, 12; **underestimate:** the figure used by

city officials—approximately 800—was calculated from surveys of industries subject to federal water pollution regulations for major indirect dischargers. But there are also an unknown number of industrial and commercial indirect dischargers that are not covered in the city's samplings. These include laundries, food establishments, car washes, gasoline stations, and other small-quantity toxic generators that often flush their wastes into city sewers with little or no government scrutiny. See Larry A. Klein *et al.*, "Sources of Metals in New York City Wastewater," *Journal of Water Pollution Control*, Vol. 46, No. 12 (December 1974), in City of New York, Department of Environmental Protection, *Industrial Pretreatment Program, Task 4, Technical Information Review* (May 1983), at Appendix C; U.S. Environmental Protection Agency, Office of Water Regulations and Standards, *Report to Congress on the Discharge of Hazardous Wastes to Publicly Owned Treatment Works* (February 1986).

25. **Zinc leads the way:** City of New York, Department of Environmental Protection, *New York City Sludge Management Plan, Task 15* (January 1990), at exhibit 14; **households and small businesses:** *Industrial Pretreatment Program, Task 4, Technical Information Review* (May 1983), at 85, Appendix C; see also Robert U. Ayres and Samuel R. Rod, "Patterns of Pollution in the Hudson-Raritan Basin," *Environment*, Vol. 28, No. 4 (May 1986), at 41-42.

26. **3,000 pounds:** City of New York, Department of Environmental Protection, *Industrial Pretreatment Program, Task 4, Technical Information Review* (May 1983), at 85-87; **organic chemicals:** J. A. Mueller and C. E. Werme, "Contaminant Inputs to the Hudson-Raritan Estuary," in Ronald J. Breteler, ed., *Chemical Pollution of the Hudson-Raritan Estuary*, U.S. Department of Commerce, National Oceanic and Atmospheric Administration (December 1984), at 128.

27. 33 U.S.C. §1342.

28. 33 U.S.C. §1317(b); 40 C.F.R. §403.2; 40 C.F.R. §§403.6, 405-471; 40 C.F.R. §403.5.

29. 33 U.S.C. §1329; 33 U.S.C. §1342(p) (4) (A); 33 U.S.C. §1342(p) (5) (B).

30. 33 U.S.C. §1251(a) (3).

31. **Federally approved program:** New York City was given full authority to run its industrial pretreatment program in January 1987. Letter from Christopher J. Daggett, Regional Administrator, U.S. Environmental Protection Agency, Region II, to Harvey W. Schultz, Commissioner, City of New York, Department of Environmental Protection (January 26, 1987). Before the passage of the Federal Water Pollution Control Act in 1972, the city had established a small industrial wastes control unit to enforce its own locally adopted sewer ordinances. See Larry A. Klein, "Control of Industrial Wastewater by New York City in the Twentieth Century," *Municipal Engineers Journal*, Vol. 70 (1984), at 17-34.

32. City of New York, Department of Environmental Protection, *Sludge Management Plan, New York City Industrial Pretreatment Program* (October 1989), at exhibits 8 and 9.

33. Letter from Paul J. Molinari, Chief, Water Permits and Compliance Branch, U.S. Environmental Protection Agency, Region II, to Edward Wagner, Assis-

tant Commissioner, New York City Department of Environmental Protection (January 29, 1988).

34. See N.Y. Envtl. Conserv. Law §§0801 *et seq.* (McKinney 1984); 6 N.Y.C.R.R. §§750 *et seq.*, creating New York State's water pollution permitting program.

35. The State Department of Environmental Conservation claims that it is unable to set toxic water quality–based limits for New York City direct dischargers, in part because of a lack of data on New Jersey and urban runoff sources. New York State Department of Environmental Conservation, "Interim Procedures, New York Harbor—Wasteload Allocation Analysis" (June 1986).

36. City of New York, Department of Environmental Protection, *Citywide Combined Sewer Overflow Study*, Final Report (November 1986), at Table 5 (6.67-hour storm).

37. **Metals . . . and organic chemicals:** City of New York, Department of Environmental Protection, *Industrial Pretreatment Program, Task 4, Technical Information Review* (May 1983), at 195-201; City of New York, Department of Environmental Protection, *Industrial Pretreatment Program, Task 7, Existing Pollutant Removals—Volume 1* (March 1984), at 3-14–3-16; see also Thomas L. Sieger and John T. Tanacredi, "Contribution of Polynuclear Aromatic Hydrocarbons to Jamaica Bay Ecosystem Attributable to Municipal Wastewater Effluents," in *Proceedings of the Second Conference on Scientific Research in the National Parks*, San Francisco (1979); **leachate:** see, generally, U.S. Department of Interior, National Park Service, Gateway National Recreation Area, Office of Resource Management and Compliance, *Summary of Several Investigations Regarding the Natural Resources of Jamaica Bay and Their Relationship to Environmental Contaminants* (March 1987); Robert C. Ahlert *et al.*, *Evaluation of Landfill Closure Requirements*, U.S. Department of Interior, National Park Service, Gateway National Recreation Area (November 1985), at v-x, 37-41; David R. Franz and William H. Harris, *Final Report: Benthos Study, Jamaica Bay Wildlife Refuge, Gateway National Recreation Area, Brooklyn, New York* (1985), at 3, 47-48.

38. **Shellfish harvesting:** 6 N.Y.C.R.R. §§41.1, 41.2; **heavy metals and organic pollutants:** Robert C. Ahlert *et al.*, *Evaluation of Landfill Closure Requirements*, U.S. Department of Interior, National Park Service, Gateway National Recreation Area (November 1985), at 37-39; the consulting firm of Parsons Brinckerhoff-Cosulich, *Investigation of Indicator Pollutant Levels at New York City Landfills* (June 2, 1982), at 23, 24, 35; **contaminants . . . turn up:** see, generally, David R. Franz and William H. Harris, *Final Report: Benthos Study, Jamaica Bay Wildlife Refuge, Gateway National Recreation Area, Brooklyn, New York* (1985); U.S. Department of Interior, National Park Service, Gateway National Recreation Area, Office of Resource Management and Compliance, *Summary of Several Investigations Regarding the Natural Resources of Jamaica Bay and Their Relationship to Environmental Contaminants* (March 1987), at 2, Section XII.

39. The Trust for Public Land with the New York City Audubon Society, *Buffer the Bay: A Survey of Jamaica Bay's Unprotected Open Shoreline and Uplands* (March 1987), at 6.

40. **33,000 new residents and employees:** The Trump Organization, *Trump City Project, Draft Environmental Impact Statement*, Vol. 1 (June 1988), at

III. B-1; **2.3 million gallons of sewage:** *Id.*, Vol. 2 (May 1988), at III. M-8; **near or at its design capacity:** New York City Department of Environmental Protection, *Monthly Operating Efficiency* (January 1989-June 1989); **7,300-car parking facility:** *Trump City Project, Draft Environmental Impact Statement, id.* at II. A-19; **25,000 vehicles:** personal communication, Daniel Gutman, air quality consultant to Westpride, June 15, 1989; **23,000 new subway trips:** *Trump City Project, Draft Environmental Impact Statement,* Vol. 2 at III. F-132; **shadow:** *Id.*, Vol. 2 at III. B-56, III. B-179, III. B-180B, III. B-180.

41. **Properties . . . targeted:** New York City Public Development Corporation, *New York City's Waterfront: A Plan for Development* (July 1986), at 29, 46, 49; **conservationists argue:** The Trust for Public Land with the New York City Audubon Society, *Buffer the Bay: A Survey of Jamaica Bay's Unprotected Open Shoreline and Uplands* (March 1987).

42. **Coastal Zone Management Act:** 16 U.S.C. §§1451 *et seq.;* **to cash in:** 16 U.S.C. §1454.

43. **Enabling legislation:** N.Y. Exec. Law §§910 *et seq.* (McKinney 1988); **44 broad policy objectives:** U.S. Department of Commerce, National Oceanic and Atmospheric Administration Office of Coastal Zone Management, *State of New York Coastal Management Program and Final Environmental Impact Statement* (August 1982); **New York City:** *Id.* Vol. 3, Appendix G, New York City Waterfront Revitalization Program (August 1982).

44. **Clean Water Act:** 33 U.S.C. §1344; **Rivers and Harbors:** 33 U.S.C. §403; **federal and state law:** National Environmental Policy Act (NEPA), 42 U.S.C. §§4321 *et seq.;* and see State Environmental Quality Review Act (SEQRA), N.Y. Envtl. Conserv. Law §§8-0101 *et seq.* (McKinney 1984). The City of New York has its own environmental review process. See Mayoral Executive Order No. 91 (August 24, 1977); **explicitly find:** N.Y. Envtl. Conserv. Law §8-0109(8) (McKinney 1984).

45. **Identify and map:** N.Y. Envtl. Conserv. Law §§34-0101 *et seq.* (McKinney 1984); **saltwater and freshwater:** N.Y. Envtl. Conserv. Law §§24-0101 *et seq.* (Freshwater Wetlands Act) and §§25-0101 *et seq.* (Tidal Wetlands Act) (McKinney 1984).

46. New York City Public Development Corporation, *New York City's Waterfront: A Plan for Development* (July 1986).

47. New York State Department of Environmental Conservation, *Investigation: Sources of Beach Washups in 1988* (December 1988), at 1-2.

48. U.S. Environmental Protection Agency, Region II, "Ocean Fact Sheet Package" (April 1988); U.S. Environmental Protection Agency, Region II, Marine Dumping Program, Ocean Dumping Program, *Quarterly Report,* October 1, 1987 to December 31, 1987; personal communication with Douglas Pabst, U.S. Environmental Protection Agency, Region II (April 7, 1989).

49. **12-Mile . . . Site:** 50 Fed. Reg. 14336, 14340 (April 11, 1985); **dredge disposal:** U.S. Environmental Protection Agency, "Ocean Dumping: Final Designation of Site," 49 Fed. Reg. 19012, 19016 (May 4, 1984); U.S. Environmental Protection Agency, *Environmental Impact Statement for the New York Dredged Material Disposal Site Designation* (August 1982), at 4-7–4-11;

U.S. Congress, Office of Technology Assessment, *Wastes in Marine Environments* (April 1987), at 65; **106-mile . . . site:** U.S. Environmental Protection Agency, Region II, "Ocean Fact Sheet Package" (April 1988), at 3; "Shell Bacteria Kill East Coast Lobsters Near Ocean Dump," *New York Times* (May 22, 1988).

50. **Ocean Dumping Act:** 33 U.S.C. §§1401 *et seq.;* 33 U.S.C. §1412(a); 40 C.F.R. Part 227; 33 U.S.C. §1413; **1988 amendments:** 33 U.S.C. §1414(b).

Part Three: Air Pollution

1. For two strong summaries of air pollution issues generally, see National Tuberculosis and Respiratory Disease Association, *Air Pollution Primer* (New York, 1971) and American Lung Association of New Jersey, *Air Pollution in New Jersey: Problems, Programs, Progress* (New Jersey, 1979, 1988).

   For a general introduction to the science of air pollution, see Charles E. Kupchella and Margaret C. Hyland, *Environmental Science: Living Within the System of Nature,* at 280-385 (Boston: Allyn & Bacon, 1986, 1989).

2. *Ethyl Corp. v. EPA,* 541 F.2d 1, 13 (D.C. Cir) *(en banc), cert. denied,* 426 U.S. 941 (1976).

3. U.S. Environmental Protection Agency, Office of Air and Radiation, "Indoor Air Facts No. 5: Environmental Tobacco Smoke" (June 1989).

4. **390,000 Americans** (and an excellent summary of the health consequences of smoking): U.S. Department of Health and Human Services, Centers for Disease Control, *Morbidity and Mortality Weekly Report,* "The Surgeon General's 1989 Report on Reducing the Health Consequences of Smoking: 25 Years of Progress (Executive Summary)," Vol. 38, No. S2 (March 24, 1989), at 9; **11,000 New York City residents:** New York Lung Association, "What Every New Yorker Ought to Know About a Smoke-Filled Room."

5. J. L. Repace and A. H. Lowrey, 1985. *A Quantitative Estimate of Nonsmokers' Lung Cancer Risk from Passive Smoking,* Environ. Internat., Vol. 11 (1985), at 3-22.

6. **Dropping by half a percent:** U.S. Department of Health and Human Services, Centers for disease Control, *Morbidity and Mortality Weekly Report,* "Tobacco Use by Adults—United States, 1987," Vol. 38, No. 40 (October 13, 1989); **citywide figure:** telephone conversation with Helen Bzduch, Research Scientist, Bureau of Adult and Gerontological Health, New York State Department of Health (October 16, 1989).

7. **New York City's statute:** New York City Administrative Code, Title 17, Ch. 5, §§1-14; **New York State's Clean Indoor Air Act:** Public Health Law, Ch. 244, Article 13E, §1300N-1399X.

8. **Total miles traveled:** Telephone conversation with Boris Pushkarev, Senior Vice President, Regional Plan Association (November 14, 1989); **760,000 cars . . . 100,000 more:** New York Metropolitan Transportation Council, *Hub-Bound Travel 1988,* Tri-State Regional Planning Commission, *Hub Bound Travel 1980.*

9. Clean Air Act Amendments of 1977, Section 202; 42 U.S.C. §§7401, 7521.
10. New York City Air Pollution Control Code; Administrative Code of the City of New York §1403.2-11.15 (vehicle idling); 1403.2-9.05 (visible emissions).
11. Natural Resources Defense Council, *The Diesel Problem in New York City* (New York, 1985), at 28. See also U.S. Department of Health and Human Services, National Institute for Occupational Safety and Health, "NIOSH Current Intelligence Bulletin 50: Carcinogenic Effects of Exposure to Diesel Exhaust" (Washington, D.C., August 1988).
12. 49 *Fed. Reg.* 40258, 40260 (October 15, 1984).
13. 46 *Fed. Reg.* 34053 (June 30, 1981).
14. National Academy of Sciences, National Research Council, *Diesel Cars: Benefits, Risks, and Public Policy* (Washington, D.C., 1982), at 111.
15. 50 *Fed. Reg.* 10620 (March 15, 1985).
16. Telephone conversation with Eugene Tierney, U.S. Environmental Protection Agency, Office of Mobile Sources, Ann Arbor, Michigan (November 28, 1989), and Michael P. Walsh, Consultant, Arlington, Virginia (November 28, 1989).

   These figures were calculated based on average emission rates for motor vehicles in operation during 1989 as derived from U.S. Environmental Protection Agency's Mobile 4 Emissions Model (using standard defaults). The model formulates CO emissions of 29.514 grams/mile, hydrocarbon emissions of 4.750 grams/mile, and NO_x emissions of 2.436 grams/mile.

   The figure for particulate emissions is based on lifetime averages for 1989 model autos, using a similar EPA emissions model and extrapolating upward for motor vehicles in operation during 1989. The model derives particulate emissions of 0.5 grams/mile.

   The figure for CO_2 emissions is based on an EPA approximation for light-duty vehicles with corporate average fuel economy of 27.5 miles/gallon, and derives CO_2 emissions of 398.1 grams/mile.

   All calculations above assume 10,000 miles traveled per vehicle per year.
17. Parsons Brinckerhoff Quade & Douglas, Inc., New York City Taxi and Limousine Commission, *Additional Taxicab Licenses: Final Environmental Impact Statement* (June 1989), at III-43.
18. *Friends of the Earth v. Carey*, 535 F.2d 165, 172 (2d Cir. 1976).
19. New York State Department of Environmental Conservation, "New York Metroplitan Area VOC Emissions Inventory" (February 10, 1989).
20. Barry Commoner, "A Reporter at Large: The Environment," *New Yorker* (June 15, 1987), at 46 (reprinted in Peter Borrelli, *Crossroads: Environmental Priorities for the Future* (Washington, D.C.: Island Press, 1987).
21. **2,700 nationwide cancer deaths:** U.S. Environmental Protection Agency, Office of Air Quality Planning and Standards, *Cancer Risk from Outdoor Exposure to Air Toxics* (external review draft, September 1989); **cancer risks . . . from 1 in 10,000 to 1 in 1,000:** U.S. Environmental Protection Agency, "Analysis of Air Toxics Emissions, Exposures, Cancer Risks and Controllability in Five Urban Areas" (Research Triangle Park, 1989), at iii.
22. *Id.* at 42, 45, and 46.
23. Telephone conversation with Jason Grumet, New York State Department of Environmental Conservation (Albany, New York) (November 29, 1989).

24. The vapor recovery strategy and three other ozone control measures were implemented by the state under court order after a Clean Air Act citizens' lawsuit was brought by New York environmental groups. See *Natural Resources Defense Council v. New York State Department of Environmental Conservation*, 668 F. Supp. 848 (S.D.N.Y. 1987).

25. James S. Cannon, INFORM, *Drive for Clean Air: Natural Gas and Methanol Vehicles* (New York, 1989).

26. **Highest respiratory cancer death rates:** New York City Department of Health, Bureau of Health Statistics and Analysis, "Summary of Vital Statistics 1987, The City of New York" (1987), at Table 3; **air pollution complaints:** Interstate Sanitation Commission, *Annual Report* (1988).

27. Telephone interview with Dr. John Oppenheimer, Director, Center for Environmental Science, College of Staten Island (November 18, 1989).

28. New York City Department of Finance, Real Property Assessment Bureau, *Recapitulation of Building Classifications* (June 1989).

29. New York City Air Pollution Control Code, Administrative Code of the City of New York §1403.2-13.03.

30. **Con Edison . . . facilities:** Telephone conversation with Dominic Mormile, Consolidated Edison spokesman (November 20, 1989); **sulfur dioxide from all sources:** New York State Department of Environmental Conservation, "1985 NAPAP Emission Inventory: Total New York City Emissions, Point and Area Sources."

31. Telephone conversation with Samuel Lieblich, New York State Department of Environmental Conservation (New York City) (November 14, 1989).

32. U.S. Environmental Protection Agency, Region II, Air and Waste Management Division, "Report on Emissions Inventory Development for Volatile Organic Compounds from Publicly Owned Treatment Works in the New York and New Jersey Ozone Nonattainment Area" (October 1987).

33. Telephone conversation with Matthew McCarthy, U.S. Environmental Protection Agency, Region II (November 21, 1989) (figures quoted from New York State emissions inventory for 1987, aircraft emissions for New York City metropolitan area).

34. American Lung Association of New York State, New York Environmental Institute, Inc., "Air Toxics in New York State: A Citizen's Guide to the Right-to-Know Law & Air Toxic Data" (Albany, 1989), at Table I.

35. U.S. Environmental Protection Agency, "AP42 Emission Factors" (Washington, D.C., April 1981), at Table 4.1-1.

36. Community Environmental Health Center at Hunter College, "Hazardous Neighbor? Living Next Door to Industry in Greenpoint-Williamsburg" (New York, 1989).

37. **Two to five times:** U.S. Environmental Protection Agency, Office of Air and Radiation, "Indoor Air Facts No. 1: EPA and Indoor Air Quality" (June 1987); **comparative risk assessment:** U.S. Environmental Protection Agency, "Unfinished Business: A Comparative Assessment of Environmental Problems" (February 1987).

38. **Second leading cause:** New York State Health Department, "Radon Update" (October 1988), Vol. 1, No. 1; **5,000 to 20,000:** U.S. Environmental Protection Agency, Office of Air and Radiation and U.S. Department of Health and

Human Services, Centers for Disease Control, "A Citizens' Guide to Radon: What It Is and What to Do About It" (August 1986) (OPA–86–004).

39. **1 in 10:** U.S. Environmental Protection Agency, Office of Public Affairs, Environmental News Release/Fact Sheet on radon action program (August 4, 1987); **1 in 20:** New York State Health Department, "Radon Update" (October 1988), Vol. 1, No. 1.

40. Anthony V. Nero, Jr., "Controlling Indoor Air Pollution," *Scientific American*, Vol. 258, No. 5 (May 1988), at 42; Congressional Research Service, The Library of Congress, by Mira Courpas, Environment and Natural Resources Policy Division; Christopher H. Dodge and Fred J. Sissine, Science Policy Research Division; Issue Brief: "Indoor Air Pollution Updated August 16, 1988," Vol. XI, No. 33 (September 2, 1988).

41. **"Action level":** U.S. Environmental Protection Agency, Office of Radiation Programs, "Radon Reference Manual" (Washington, D.C., September 1987); **Indoor Radon Abatement Act:** Indoor Radon Abatement Act, Pub. L. No. 100-552, 102 Stat. 2755 (1988).

42. National Academy of Sciences, *National Issues in Science and Technology: Global Environment Change* (Washington, D.C., 1989), at 58.

43. Ted R. Mill *et al.*, The Urban Institute, "Global Climate Change: A Challenge to Urban Infrastructure: Abstract" (Washington, D.C., 1989), at 8-10.

Part Four: Drinking Water

1. For a detailed history of the city's drinking water system, see Charles H. Weidner, *Water for a City: A History of New York City's Problem from the Beginning to the Delaware River System* (Rutgers University, (Quinn & Boden Co., Inc., Rahway, NJ, 1974); see also Robert Alpern, Citizens Union Foundation, *Water-Watchers: A Citizens Guide to New York City Water Supply* (1987), at 10-22. And see *New Jersey v. New York*, 283 U.S. 336 (1983); decree amended 347 U.S. 995 (1954).

2. For more detailed discussion of the city's water supply infrastructure, see New York City Municipal Water Finance Authority, "Water and Sewer System Revenue Bonds Fiscal 1989 Series B" (March 15, 1989); and Robert Alpern, Citizens Union Foundation, *Water-Watchers: A Citizens Guide to New York City Water Supply* (1987), at 23-35.

3. New York State Department of Environmental Conservation, *Delaware–Lower Hudson Region Water Resources Management Study* (September 1987), at 4-271.

4. New York State Senate Research Service, Task Force on Critical Problems, *Water Conservation: The Hidden Supply* (September 1986), at 35.

5. New York State Department of Health, "An Analysis of New York City Safe Yield Deficiencies and Recommended Action" (February 1986), at Table 1; New York City Municipal Water Finance Authority, "Water and Sewer System Revenue Bonds, Fiscal 1989 Series B" (March 15, 1989); see also Robert Alpern, Citizens Union Foundation, *Thirsty City: A Plan of Action for New York City Water Supply* (1986), at 39, Appendix B.

6. **Upstate usage:** New York State Department of Health, "An Analysis of New

York City Safe Yield Deficiencies and Recommended Action" (February 1986), at Table 1; **Dutchess County:** Cara Lee, "Drought or a Shortage? New York City's Quest for Water," in *Water for Millions: At What Cost?*, Cara Lee, ed. (Poughkeepsie, N.Y.: Scenic Hudson, 1987).

7. New York State Department of Environmental Conservation, New York State Water Resources Planning Council, *Water Resources Management Strategy* (January 1986), at 1-6.

8. 1905 N.Y. Laws 724, 725, 726.

9. 1954 ruling: *New Jersey v. New York*, 347 U.S. 995 (1954). The parties entered into a revised set of accords in 1982. See the so-called "good faith" agreement, Delaware River Basin Commission, "Recommendations of the Parties to the U.S. Supreme Court Decree of 1954, Pursuant to Commission Resolution 78-20," (November 1982).

10. **1989 city statute:** New York City Local Law No. 29 of 1989 (signed by mayor, May 15, 1989); **emergency regulations:** New York City Department of Environmental Protection, "Drought Emergency Rules" (1989).

11. Mayor's Intergovernmental Task Force on New York City Water Supply Needs, *Managing for the Present, Planning for the Future*, (Second Interim Report, December 1987); Mayor's Intergovernmental Task Force on New York City Water Supply Needs, *Increasing Supply, Controlling Demand* (Interim Report, February 11, 1986).

12. City of New York, Department of Environmental Protection, Bureau of Water Supply and Wastewater Collection, "Request for Proposal, Hudson River Project" (July 21, 1989), at 1.

13. Mayor's Intergovernmental Task Force on New York City Water Supply Needs, *Managing for the Present, Planning for the Future* (Second Interim Report, December 1987), at 16.

14. *Id.* at 44-46.

15. **60,000 such meters:** Personal communication, Steven Ostrega, assistant commissioner, New York City Department of Environmental Protection (November 16, 1989); **180,000 metered accounts:** New York City Water Board, "Report on the Metered and Flat Rate Methods of Billing Including Fiscal Year 1989 Rate Alternatives" (April 1988), at 12, Exhibit III-4; **225 to 300 million gallons a day:** Mayor's Intergovernmental Task Force on New York City Water Supply Needs, *Managing for the Present, Planning for the Future* (Second Interim Report, December 1987), at 9, Appendix A at 21; City of New York Department of Environmental Protection, "Universal Metering Program Implementation Plan, Executive Summary" (1986), at 1-5.

16. **Toilets:** Personal communication, Steven Ostrega, assistant commissioner, New York City Department of Environmental Protection (November 16, 1989); **40 percent:** New York State Senate Research Service, Task Force on Critical Problems, *Water Consumption: The Hidden Supply* (September 1986), at 35; **200 million gallons a day:** City of New York, Office of the Comptroller, *Waste Not, Want Not: Managing New York City's Water Supply* (July 1986), at 46.

17. **San José:** Personal communication, John Jamieson, Water Conservation Specialist, Office of Environmental Management, Water Resources and Conservation Program, City of San José (October 24, 1989); **20-unit building:** per-

sonal communication, Steven Ostrega, assistant commissioner, New York City Department of Environmental Protection (November 16, 1989).

18. Mayor's Intergovernmental Task Force on New York City Water Supply Needs, *Managing for the Present, Planning for the Future* (Second Interim Report, December 1987); New York City Water Board, "Report on the Metered and Flat Rate Methods of Billing Including Fiscal Year 1989 Rate Alternatives" (April 1988), at 35.

19. **Fiscal year 1988 to 1989:** City of New York, *The Mayor's Management Report* (September 17, 1989), at 160; **speeding up:** Mayor's Intergovernmental Task Force on New York City Water Supply Needs, *Managing for the Present, Planning for the Future* (Second Interim Report, December 1987), at 7.

20. Personal communication, Thomas O'Connell, Chief Strategic Services Division, Bureau of Water Supply and Wastewater Collection, New York City Department of Environmental Collection (December 28, 1989).

21. 33 U.S.C. §1342; N.Y. Envtl. Conserv. Law §§17-0801 *et seq.* (McKinney 1984).

22. **State law:** N.Y. Pub. Health Law §1100 (McKinney 1989); **watershed regulations:** The City of New York, Department of Water Supply, Gas and Electricity, "Rules and Regulations for the Protection from Contamination of the New York City Water Supply and Its Sources" (August 10, 1953).

23. 42 U.S.C. §300g-1.

24. 54 Fed. Reg. 27486, 27527-29 (June 29, 1989).

25. "Water, Water Everywhere," *Consumer Reports* (January 1987), at 42.

26. **Trihalomethanes:** See, for example, City of New York, Department of Environmental Protection, *New York City Drinking Water Quality Report for the Month of January 1989* (March 2, 1989), at 41; **alum:** see, for example, affidavit of Raul R. Cardenas Jr., dated July 2, 1989, in *Hudson River Fishermen's Association v. The City of New York*, No. 89-2387 (S.D.N.Y.).

27. Personal communication with Arthur Ashendorff, Director, Bureau of Public Health Engineering, New York City Department of Health (May 5, 1989); The Plumbing Foundation City of New York, Inc., "Results of Random Testing for Lead Contamination of New York City's Water, June–August 1988" (November 16, 1988).

28. **20 wells closed:** Personal communication with Carl Becker, Vice President for Engineering, Jamaica Water Supply Company (October 27, 1989); **organic chemicals:** see, for example, New York City Department of Health, "Jamaica Water Supply Company Wells—Sampling and Wellfield Survey 1985" (December 1986), at 10.

29. Robert Alpern, Citizens Union Foundation, *Water-Watchers: A Citizens Guide to New York City Water Supply* (1987), at 25-26.

30. N.Y. Gen. City Law §20(2) (c) (McKinney 1989).

31. Figures provided by the Regional Plan Association (October 25, 1989).

32. **83 sewage treatment plants:** City of New York, Department of Environmental Protection, "New York City's Long-Range Water Quality and Watershed Protection Program" (June 1989), at 2; **10 million gallons a day:** City of New York, Department of Environmental Protection, "Watershed Sewage Treatment Plant Noncompliance Summary Reports, January–June 1987 and 1988" (December 1988).

33. New York City Department of Environmental Protection, Bureau of Water Supply and Wastewater Collection, "Watershed Sewage Treatment Plant Priority Problem List" (November 1988).
34. **30 percent of the wastewater flow:** City of New York, Department of Environmental Protection, "New York City's Long-Range Water Quality and Watershed Protection Program" (June 1989), at 2; **86 percent noncompliance:** City of New York, Department of Environmental Protection, "Watershed Sewage Treatment Plant Noncompliance Summary Report, January–June 1987 and 1988" (December 1988).
35. Memorandum from Harvey W. Schultz, Commissioner, New York City Department of Environmental Protection, to Mayor Edward I. Koch (September 16, 1988), at 2; see also New York State Department of Environmental Conservation, *Nonpoint Source Control of Phosphorous: A Watershed Evaluation, Volume 5, The Eutrophication of the Cannonsville Reservoir* (April 1986).
36. Jane Ceraso *et al.*, Environmental Defense Fund, *New York City's Water Supply: Acid Deposition, Inorganic Pollution and the Catskill Reservoirs* (October 1986), at 2-3.
37. Mark McIntyre, "New York City Drinking Water: Trouble on Tap," *New York Newsday* (August 21, 1988).
38. City of New York, Department of Environmental Protection, "New York City's Long-Range Water Quality Watershed Protection Program" (June 1989).
39. The New York Academy of Medicine, Committee on Public Health, "Statement on Preservation of New York City's Drinking Water Quality" (June 14, 1989).
40. **Independent water tasters:** *Consumer Reports.* "The Selling of H_2O" (September 1980), at 531 and "Water, Water Everywhere," *Consumer Reports* (January 1987), at 42.
41. **Suffolk County survey:** Suffolk County Department of Health Services, "Water Quality Survey of Bottled Water and Bottled Water Substitutes" (January 1988), at 3. See also New York State Department of Health, Bureau of Public Water Supply Protection, "Survey of Volatile Organic Chemical Compounds in Bottled Water Products Distributed in New York State" (January 1987); **a third of all bottled waters:** Environmental Policy Institute, "Bottled Water: Sparkling Hype at a Premium Price" (Washington, D.C. January 1989), at 1.

Part Five: Toxics

1. Jack Lewis, "Lead Poisoning: A Historical Perspective," *EPA Journal*, Vol. II, No. 4 (May 1985), at 15, 16.
2. See, generally, National Research Council, National Academy of Sciences, *Lead in the Human Environment*, Washington, D.C. (1980), at 38; and see U.S. Department of Health and Human Services, Public Health Service, *The*

Nature and Extent of Lead Poisoning in Children in the United States: A Report to Congress (July 1988), at I-41.

3. *Id.* at I-1.

4. New York City Department of Health, Bureau of Lead Poisoning Control, *Annual Report—1985* at 12; **city screening programs:** New York City, Mayor's Office of Operations, *The Mayor's Management Report* (September 15, 1988), at 398, 406; **testing net:** see New York City Comptroller's Office, *A Silent Epidemic: Childhood Lead Poisoning* (September 1986), at 30-35; **jump by ten times:** personal communication with Dr. John F. Rosen, Professor of Pediatrics, Albert Einstein College of Medicine, Bronx, New York (June 28, 1989).

5. **Household paints:** 42 Fed. Reg. 44192, 44198 (September 1, 1977); **single paint chip:** E. Groth, Consumers Union of the United States, Inc., "The Child in the Leaded Environment," (1981), at 2.

6. New York City Department of Health, Bureau of Lead Poisoning Control, *Annual Report—1985*, at 13, "Lead Poisoning Cases by Health District, FY 86, 87, 88" (transmitted to NRDC December 29, 1989).

7. **Congress banned:** 42 U.S.C. §4831(a), (b) and (c); **Consumer Product Safety Commission:** the action by the commission was in part a response to an earlier petition filed by Consumers Union. Among the exceptions from the ban are coatings for agricultural and industrial equipment, traffic and safety markings, graphic art coatings, artists' paints, and touch-up coatings for agricultural equipment, lawn and garden equipment and appliances. 42 Fed. Reg. 44192, 44198 (September 1, 1977).

8. **1982 city mandate:** Administrative Code of the City of New York §27-2013(h); **landlords must remove:** New York City Health Code §173.13. Under the regulations, city agencies are directed to step in and remediate the problem when property owners have failed to act within five days.

9. **27 million older housing units:** 49 Fed. Reg. 19218 (May 4, 1984); **slow to enforce:** see, for example, *Ashton v. Pierce*, 716 F.2d 56 (D.C. Cir., 1983); **federal funding:** federal funds previously earmarked for lead-screening activities have now been merged into state-administered block grant programs, in which a large number of health-related services must compete for limited funding. See U.S. Department of Health and Human Services, *supra* note 2.

10. **Number of children screened:** New York City Department of Health, Bureau of Lead Poisoning Control, *Annual Report—1985*, at 12; New York City, Mayor's Office of Operations, *The Mayor's Management Report* (Preliminary, February 15, 1989), at 315; **lead paint hazards:** see *New York City Coalition to End Lead Poisoning, et al. v. Koch, ete al.*, slip op. at 8 (N.Y. Sup. Ct., January 20, 1987). See also New York City, Office of the Comptroller, "A Silent Epidemic: Childhood Lead Poisoning" (September 1986), at 22-28.

11. **Levels at pumping stations:** See, for example, New York City Department of Environmental Protection, Division of Drinking Water Quality Control, *New York City Drinking Water Quality Report* (January 1989); **first-draw water:** personal communication with Arthur Ashendorff, Director, Bureau of Public Health Engineering, New York City Department of Health (May 5, 1989) (results based on roughly 150 samples taken over a 12-month period); **1988 survey:** the Plumbing Foundation of the City of New York, Inc., "Results of

Random Testing for Lead Contamination of New York City's Water, June–August 1988" (November 16, 1988).

12. 42 U.S.C. §300g-6; 50 Fed. Reg. 46936 (November 13, 1985).

13. **Largest source:** See, for example, U.S. Environmental Protection Agency, Office of Research and Development, "Air Quality Criteria for Lead" (December 1977), at 5-2; **3,000 tons a year:** New York State Department of Environmental Conservation, "Air Quality Implementation Plan: Lead" (June 1979), at 2, II-13.

14. **1971 city law:** N.Y.C. Admin. Code §1403.2-13.11; **phasedown program:** 38 Fed. Reg. 33734-41 (December 6, 1973). See also *Exxon Corp. v. City of New York,* 548 F.2d 1088 (2d Cir. 1977). The EPA rules were promulgated pursuant to congressional directive in the federal Clean Air Act. See 42 U.S.C. §7545. That same year, EPA also directed that lead-free gasoline be made available at service stations beginning in 1975. 38 Fed. Reg. 1254 (January 10, 1973); **declined by more than 95 percent:** existing federal regulations limit allowable levels of lead in gasoline to 0.1 grams per leaded gallon. 50 Fed. Reg. 9386 (March 7, 1985). Gasoline lead levels as late as 1970 were typically more than two grams per gallon.

15. A review of 1970–76 data revealed blood lead levels here decreased in virtual lockstep with government lead phasedown programs. See I. H. Billick, A. S. Currau, D. R. Shier, "Analysis of Pediatric Blood Lead Levels in New York City for 1970–76," *Environmental Health Perspectives,* Vol. 31 (1979). This trend continued after 1976 as well, according to a definitive nationwide analysis by the National Center for Health Statistics and the federal Centers for Disease Control. See U.S. Environmental Protection Agency, "Costs and Benefits of Reducing Lead in Gasoline: Final Regulatory Impact Analysis" (February 1985), at III-6–III-8.

16. 48 Fed. Reg. 56407 (December 21, 1983).

17. **Lung cancer:** See, for example, J. C. McDonald *et al.,* "Mortality in Canadian Miners and Millers Exposed to Chrysatile," *Annual New York Academy of Sciences* Vol. 330 (1979), at 1-9; I. Selikoff *et al.,* "Asbestos Disease in United States Shipyards," *Annual New York Academy of Sciences,* Vol. 330 (1979), at 295-311. Some evidence also suggests a link between asbestos exposure and cancers of the larynx, gastrointestinal tract, and kidney. See I. Selikoff *et al.,* "Mortality Experience of Insulation Workers in the United States and Canada, 1943–1976," *Annual New York Academy of Sciences,* Vol. 330 (1979), at 91-116; **between 3,300 and 12,000 cancer cases:** 51 Fed. Reg. 3738 (January 29, 1986); **asbestosis:** see National Research Council, *Asbestiform Fibers: Nonoccupational Health Risks* (1984), at 112-14; **cigarette smoking:** for example, the risk of lung cancer is five times greater for males who smoke and work with asbestos than it is for smokers who are not exposed to this substance. E. C. Hammond *et al.,* "Asbestos Exposure, Cigarette Smoking and Death Rates," *Annual New York Academy of Sciences,* Vol. 330 (1979), at 473-90.

18. **One-time acute dose:** National Institute for Occupational Safety and Health, "Criteria for a Recommended Standard: Occupational Exposure to Asbestos" (1972); **no known safe level:** National Institute for Occupational Safety and Health, Safety and Health Administration Asbestos Work Group, Workplace

Exposure to Asbestos (1980), at 3; National Research Council, *Asbestiform Fibers: Nonoccupational Health Risks* (1984), at 11.

19. See A. N. Rohl *et al.*, "Asbestos Content of Dust Encountered in Brake Maintenance and Repair," *Proceedings of the Royal Society of Medicine*, Vol. 70 (1977). See also A. N. Rohl *et al.*, "Airborne Asbestos in the Vicinity of a Freeway," *Atmospheric Environment*, Vol. 12 (1978).

20. New York City Department of Environmental Protection, *Final Report on the Assessment of the Public's Risk of Exposure to In-Place Asbestos* (December 1, 1988), at xvi, xvii, 2-21.

21. New York City Board of Education, "Asbestos Abatement Program Information Update Report" (December 1985).

22. See 42 U.S.C. §§7401 *et seq.* (Clean Air Act); 15 U.S.C. §§2901 *et seq.* (Toxic Substances Control Act); 29 U.S.C. §§681 *et seq.* (Occupational Safety and Health Act).

    Three separate statutes have been enacted by Congress to address the asbestos-in-schools problem: (1) 20 U.S.C. §§3601 *et seq.* (the 1980 Asbestos School Hazard Detection and Control Act); (2) 20 U.S.C. §§4011 *et seq.* (the 1984 Asbestos School Hazard Abatement Act); and (3) 15 U.S.C. §§2641–2654 (the 1986 Asbestos Hazard Emergency Response Act).

23. **Spray-on form:** 40 CFR 61.148, 61.150; **protect workers:** in a series of regulatory actions since the early 1970s, the United States Occupational Safety and Health Administration has set maximum permissible exposure limits for airborne asbestos in the workplace. Under new rules promulgated in 1986, the asbestos limit has been cut by more than 90 percent from occupational levels set a decade ago. Still, the National Institute for Occupational Safety and Health has recommended a workplace limit one-half the current standard. 51 Fed. Reg. 22,612, 22,614 (June 20, 1986); **phase out 94 percent:** 54 Fed. Reg. 29460 (July 12, 1989).

24. **School systems to survey:** 52 Fed. Reg. 41 826 (October 30, 1987); **licensing of . . . contractors:** N.Y. Lab. Law §§900 *et seq.* (McKinney 1988); **encapsulation:** N.Y. Lab. Law §906 (McKinney 1988).

25. 13 N.Y.C.R.R. §§18.7(aa), 20.7(z), 21.7(z), 23.7(cc).

26. **Spray-on asbestos:** Administrative Code of the City of New York, §24-146(b); **in-place asbestos:** Administrative Code of the City of New York, §24-146.1, §27-198.1; **storage, transport, and disposal:** Administrative Code of the City of New York, §16-117.1.

27. New York City Department of Environmental Protection, "Asbestos Control Program, Table of Penalties" (November 1988).

28. Personal communication with Frederic L. Sachs, Director, Asbestos Control Program, New York City Department of Environmental Protection (May 5, 1989).

29. See Leonard Buder, "25 Charged in Bribes Tied to Asbestos," *New York Times* (January 6, 1988); and see Pete Bowles, "Wrecker Convicted in EPA Bribery Case," *New York Newsday* (November 30, 1988).

30. U.S. Environmental Protection Agency, Office of Toxic Substances, *Support Document/Voluntary Environmental Impact Statement for Polychlorinated Biphenyls (PCBs) Manufacturing, Processing, Distribution in Commerce, and Use Ban Regulation (Section 6(e) of TSCA)* (April 1979), at 2.

31. See *Id.* at 18; 51 Fed. Reg. 28556, 28558-59 (August 8, 1986); 50 Fed. Reg. 29170-71 (July 17, 1985); and U.S. Environmental Protection Agency, Office of Toxic Substances, "The PCB Regulation Under TSCA: Over 100 Questions and Answers to Help You Meet These Requirements" (August 1983), at 10.

32. U.S. Environmental Protection Agency, Region II, New York State Department of Environmental Conservation, "Draft Joint Supplement to the Final Environmental Impact Statement on the Hudson River PCB Reclamation Demonstration Project" (January 1987), Appendix F at i.

33. **Most severely PCB-contaminated:** U.S. Environmental Protection Agency, Region II, New York State Department of Environmental Conservation, "Draft Joint Supplement to the Final Environmental Impact Statement on the Hudson River PCB Reclamation Demonstration Project" (January 1987), Appendix F at i; **20 species of fish:** *Id.*, Appendix F at 39-40; **two or more times federal standards:** New York State Department of Environmental Conservation, *PCB Concentrations in the Striped Bass from the Marine District of New York State* (April 1988), at Table 5, Table 25; New York State Department of Environmental Conservation, *PCB in Hudson River Striped Bass: Ten Years of Monitoring* (June 1988), at Table 3; **nine species off limits:** 6 N.Y.C.R.R. §§11.1, 11.2, 11.3; New York State Department of Health, *1987– 1988 Health Advisory* (July 13, 1987); **economic losses:** see Prepared Testimony of Bruce D. Shupp, Chief, Bureau of Fisheries, New York State Department of Environmental Conservation, "In the Matter of the Application of the PCB PROJECT GROUP for Approvals of the HUDSON RIVER PCB RECLAMATIONS DEMONSTRATION PROJECT" (June 30, 1987), at 16.

34. *New York Times*, "PCB-Laden Offices to Be Ventilated" (February 10, 1985); personal communication with David Rings, Director, Occupational Safety and Health, New York State Office of General Services (May 11, 1989).

35. **1,440 large PCB transformers:** Letter from Carol B. Hafer, Counsel, New York City Fire Department, to NRDC (March 9, 1987); **PCB spills:** U.S. Coast Guard, National Response Center, database for all PCB-related incidences in New York State (January 1986 to January 1989).

36. **Toxic Substances Control Act:** 15 U.S.C. §2601, 2605(e); **1979 . . . regulations:** 44 Fed. Reg. 31514 (May 31, 1979); **existing transformers:** 50 Fed. Reg. 29170 (July 17, 1985); **Con Edison, and others:** personal communication, Larry Gerschwer, Division of Hazardous Materials Program, New York City Department of Environmental Protection (May 12, 1989); personal communication, Robert Keegan, Director, Water and Waste Management Program, Consolidated Edison Company (May 16, 1989); see also New York City Department of Environmental Protection, *PCB Transformer Program* (April 1, 1985).

37. **Community right-to-know:** 42 U.S.C. §§11001-11050; **submit information:** 42 U.S.C. §11023; **disclose . . . chemicals:** 42 U.S.C. §§11002, 11021-11022; **develop emergency response:** 42 U.S.C. §§11001–11003.

38. Administrative Code of the City of New York, §§24–701 *et seq.*

39. **Filed disclosure forms:** New York State Department of Environmental Conservation, "SARA Title III 1987 Toxic Chemical Release Inventory Data" (October 19, 1988); New York City Department of Environmental Protection, Division of Hazardous Materials, Right-to-Know Section, "Right-to-Know

List" (April 5, 1989); **stimulated . . . emergency planning:** New York City, Mayor's Emergency Control Board, *Superfund Amendments and Reauthorization Act (S.A.R.A.) Title III Plan* (1988).

40. In a 1988 report, New York State Department of Environmental Conservation calculated that the city's large and mid-size generators had produced nearly 50,000 tons of hazardous waste in the previous year. New York State Department of Environmental Conservation, *1987 Report on the Generation of Hazardous Waste in New York State* (December 1988), at 529. Compare also New York State Department of Environmental Conservation, *Small Quantity Generators in New York State, Draft Final Report* (December 1984), at 27.

41. **13,000 may be operating:** New York State Department of Environmental Conservation, *Small Quantity Generators in New York State, Draft Final Report* (December 1984), at 14; **number at about 70:** personal communication with Ernest Robbins, Division of Hazardous Substance Regulation, New York State Department of Environmental Conservation (May 9, 1989). Figure based on those firms generating more than roughly 14 tons per year (1,000 kilograms per month) of hazardous waste. Data derived from New York State Department of Environmental Conservation, Division of Solid Hazardous Waste, *1987 Detailed Generator Annual Report* (September 24, 1988); **60 drums:** New York State Joint Legislative Commission on Toxic Substances and Hazardous Wastes, *Educating Small Quantity Generators: New Challenges for Hazardous Waste Management* (March 1989), at 1; **two largest generators:** New York State Department of Environmental Conservation, Division of Solid and Hazardous Waste, *1987 Detailed Generator Annual Report* (September 24, 1988).

42. **Common household toxics:** Environmental Hazards Management Institute, "Household Hazards Waste Wheel," (Portsmouth, New Hampshire, 1987); **$1/3$ to $1/2$ of 1 percent:** U.S. Environmental Protection Agency, Environmental Monitoring Systems Laboratory, "Characterization of Household Waste from Marin County, California, and New Orleans, Louisiana" (April 1987), at 1; U.S. Environmental Protection Agency, Office of Solid Waste and Emergency Response, *A Survey of Household Hazardous Wastes and Related Collection Programs* (October 1986), at 4-11; **100,000 pounds a day:** figures based on New York City Sanitation Department regular collection of residential waste from July to October 1989. New York City Department of Sanitation, Bureau of Waste Disposal, "Loads and Tonnage Report" (October 1989), at 3.

43. **29 city locations:** New York State Department of Environmental Conservation, *Quarterly Status Report of Inactive Hazardous Waste Disposal Sites* (October 1989), at 28-29; **reconnaissance by federal investigators:** U.S. Environmental Protection Agency, Superfund Program, CERCLA, "List-8: Site/Event Listing" (May 5, 1988).

44. **Comprehensive framework:** 42 U.S.C. §§6901 *et seq;* manifest system: 40 CFR 260 *et seq;* **New York State . . . program:** N.Y. Envtl. Conserv. Law §§27-0900 *et seq* (McKinney 1984); and 51 Fed. Reg. 17737 (May 15, 1986).

45. **Federal Superfund:** 42 U.S.C. §§9601 *et seq;* **national priority list:** 40 CFR 300.68.

46. N.Y. Envtl. Conserv. Law §§27-1301 *et seq* (McKinney 1984). The state

program is primarily funded through fees imposed on hazardous waste generators and transporters. N.Y. Envtl. Conserv. Law §27-0923 (McKinney Supp. 1989).

47. See, most recently, New York State Department of Environmental Conservation, *Quarterly Status Report of Inactive Hazardous Waste Disposal Sites* (October 1989), at 28-29; also New York State Department of Environmental Conservation, *Inactive Hazardous Disposal Sites in New York State*, Vol. 2 (December 1987).

48. Mount Sinai School of Medicine at the City University of New York, Department of Community Medicine, Environmental and Occupational Medicine, Occupational Disease in New York State: Proposal for a Statewide Network of Occupational Disease Diagnosis and Prevention Centers, Report to the New York State Legislature (February 1987), at 6.

49. *Id.* at 26.

50. See, generally, State of New York, Department of Environmental Conservation, *In the Matter of the Application of the RADIAC RESEARCH CORPORATION for a Permit to Operate a Hazardous Waste Management Facility DEC Application No. 20–86–0035, Hearing Report* (May 6, 1988) and *decision* (December 14, 1988). Andrew S. Pearlstein, Administrative Law Judge; **Invitation to disaster:** Prefiled Testimony of Dr. Richard Y. Levine, at 14.

51. See, generally, Paul Brodeur, "Annals of Radiation: The Hazards of Electromagnetic Fields, III-Video Display Terminals", *New Yorker* (June 26, 1989).

52. **"Grim history"**: Legislative History of the Occupational Safety and Health Act of 1970 (Committee Print Compiled for the Senate Committee on Labor and Public Welfare, 1971, p. iii); **adopt health standards:** 29 U.S.C. §§651 *et seq.*; **New York State law:** N.Y. Lab. Law §§27-a *et seq.* (McKinney 1986).

53. **Carved into law:** N.Y. Lab. Law §§875 *et seq.* (McKinney 1988); **less comprehensive:** 52 Fed. Reg. 31852 (August 24, 1987).

54. See, generally, Stephen G. Minter, "Prosecutor Charges: 'The Weapon Was a Dangerous Chemical,'" *Occupational Hazards* (September 1987), at 105-9.

55. **Systemic weaknesses:** U.S. Department of Labor, Office of Inspector General, "Special Review of OSHA Enforcement Activities" (Draft, 1987), at 1; **reshuffled top managers:** Bill Sternberg, "Turmoil Ravages Local OSHA Office," *Crain's New York Business* (February 24, 1986); Bill Sternberg, "OSHA Slaps New York Chief," *Crain's New York Business* (May 12, 1986).

Index

Abrams, Robert, 52, 181
Acidic gases, 37, 38
Acid rain, 37, 157–58
AIDS, 14, 15
Air pollution, 85–127
  background, 89, 92–94
  from commercial, industrial, and
    residential sources, 111–27
    airports, 120–21
    consumer solvents and paints, 117–
     18
    global warming and, 124–25
    incinerators, 115–16
    indoor pollutants, 123–27
    other sources, 121–22
    petroleum marketing, 117
    power plants and space heating,
     111–14
    sewage treatment plants, 118–20
  epilogue, 217–19
  from incinerators, 36–40
    acidic gases, 37
    dioxins, 37
    particulate matter, 36–37
    proposed "resource recovery," 38–
     40

  in use today, 37–38
  introduction, 87–88
  from landfills, 8–9
  from motor vehicles, 94–111
    alternative fuel programs, 108–109
    carbon monoxide, 101–102
    diesel-powered vehicles, 96–100,
     107
    gasoline-powered vehicles, 100–
     106
    government action, 106–11
    inspection programs, 107–108
    introduction, 94–95
    lead, 104–105, 175–78
    legal aspects of, 96–106
    ozone, 102–104
    pollution control devices, 95, 106,
     109
    reducing motor vehicle use, 109–
     11
    synergism and, 89, 105
    toxics, 105–106
  neighborhood severity of, 93–94,
    110–11
  ten major sources of, 90
  tobacco smoking, 91–93

Airports, pollution from, 120–21
Alley Pond Environmental Center, 215
Alternative fuels program, 108–109
Alum in drinking water, 151
American Federation of State, County
       and Municipal Employees, 223
American Lung Association, 121, 217
Apartment house heating systems,
       111–14
Apartment house incinerators, 32–33,
       37–38, 43, 116, 177
Army Corps of Engineers, 75, 84, 143
Arsenic, 36
Arthur Kill power plant, 114
Arthur Kill waterway, 10, 110
Asbestos, 105, 178–85
    background, 178–79
    exposure in N.Y.C., 179–80
    friable, 180, 183
    government action, 182–85
    health effects, 179
    legal aspects of, 181–82
    removal of, 181–85
    in schools, 180, 181
    steam pipes and, 178
    uses of, 178
    in the workplace, 179–80
Asbestos in New York City Homes, 221
Ashokan reservoir, 133, 135
Ash residue from incinerators, 12, 33,
       34, 40–41
Athens, Greece, 87
Audubon Society, 73, 215

Bacteria and oxygen levels in
       waterways, 48–50, 56, 65, 81, 83
Bacteria in drinking water, 148, 151
Beaches, closed, 14–16, 49
Beame, Abraham, 101–102
Bedford Hills Correctional Facility, 156
Bellamy, Carol, 114
Benzene, 8, 63, 105, 106

Benzopyrenes, 98
Big Apple Wrecking Corporation, 184
Biodegradability, 29
Board of Education, New York City,
       180, 185
Bottle bill, 19–23, 29
Bottled water, 160–61
"Bottom ash," 40
Bowery Bay sewage treatment plant, 51
Brant Point, 67
Bronx River Restoration, 215
Bronx 2000, 26–27
Brookfield Avenue landfill, 9, 17–18
Brooklyn Navy Yard incinerator
       (proposed), 33–35, 38, 41, 116
    projected emissions from, 39, 177–78
Brooklyn-Queens aquifer, 8, 138, 143–
       44, 151, 152–54, 158
Brooklyn Union Gas Company, 108
Building Code, New York City, 76

Cadmium:
    in solid waste, 8, 29, 36, 40, 41
    in water, 59, 62
Cancer:
    air pollution and, 98, 105–106
    cigarette smoking and, 91–93
    skin, 103
Cannonsville reservoir, 157, 162
Carbon dioxide, 100
Carbon monoxide, 88, 93, 94, 100, 101–
       102, 105, 106, 108, 120
Carey, Hugh, 102
Caro, Robert A., 77
Catalytic converters, 95, 106
Catskill watershed, 133–35, 141, 150,
       163
    map of, 136
Caustic soda in drinking water, 151
Center for the Biology of Natural
       Systems, 21
Centers for Disease Control, 91, 170

Central Park reservoir, 134, 135
Chelsea pumping station, 138, 142–43, 159
Chemicals in your community, 188
Chemical Waste Disposal Corporation, 201
Chlorine, 105
   in drinking water, 151, 152
Chlorofluorocarbons (CFCs), 28, 102, 219
Cholera, 148
Chromium, 18
   in water, 62, 63
Cigarette smoking, 91–93, 179
City Council of New York, 43
   air pollution and, 112
   asbestos and, 182
   recycling and, 27, 28, 30
Clean Air Act of 1970, 96, 102, 105, 106, 107, 116, 122, 123, 176, 181
Clean air plan of 1973, 101–102
Clean Indoor Air Act, 93
Clean Water Act of 1972:
   drinking water and, 149, 155
   waterways and, 48, 50, 51, 58, 59, 63, 68, 75
Coalition for a Livable West Side, 72
Coalition for the Bight, 215
Coast, the, 68–79
   background, 68–69
   development of, 69–74
   epilogue, 215–17
   government action, 76–77
   legal aspects of, 75–76
   maps of, 78–79
   rising seas and, 74
Coastal Zone Management Act of 1972, 75
Coastline, see Waterways and the coast
Combined sewer overflow (CSO), 49, 53–55, 58
Command Bus Company, 108

Commercial waste recycling, 24–26
Commoner, Barry, 21, 104
Community Environmental Health Center, 122
Comprehensive Environmental Response, Compensation and Liability Act (CERCLA) of 1980, 195
Con Edison, 112–14, 189, 192
Coney Island Beach, 49
Coney Island sewage treatment plant, 48, 51, 52
Conservation Directory, 211–12
Construction debris, 81
Consumer Product Safety Commission, 126, 172
Consumer solvents and paints, 117–18
Consumers Union, 160, 161, 220
Copper in water, 60, 62, 63, 65
Copper sulfate in drinking water, 151
Council on the Environment of New York City, 26, 214, 221
Court of Appeals, U.S., 102
"Criteria" pollutants, 105
Croton Falls reservoir, 156
Croton watershed, 133–35, 150–51, 155, 156, 159, 162–63
   map of, 136
Cuomo, Mario, 108
Curbside recycling, 24, 25
Cyanide, 10

DDT, 60
Dean and Deluca, 213
Delaware watershed, 133–35, 141, 150, 156–57, 163
   map of, 136
Department of Environmental Conservation (DEC), N.Y. State:
   air pollution and, 117, 122
   drinking water and, 142, 162
   hazardous wastes and, 193, 196

Department of Environmental
    Conservation (DEC) (cont'd)
  solid waste and, 9, 16, 40–43
  solid waste management plan, 19–20
  water pollution and, 52–53, 60–61,
    68, 74
Department of Environmental
    Protection (DEP), New York
    City, 218
  asbestos and, 182, 184
  drinking water and, 146, 161, 162,
    174, 219
  sewage pollution and, 49, 54, 55, 56
Department of General Services, New
    York City, 30
Diesel particulate matter, 88, 93, 96–
    100, 105, 107
Dinkins, David, 24, 43, 109
Dioxins, 186
  in solid waste, 17, 29, 37, 38, 41, 42
  in water, 59
Direct dischargers, 60–61
  Clean Water Act and, 63
  government action on, 68
Dredge material, 81, 83
Drinking water quality, 148–63
  additives, 151
  background, 148
  bottled water, 160–61
  differences in, 150–51
  end-of-the-line plumbing and, 155
  epilogue, 219–21
  filtration and, 149–50, 158–59, 161,
    162–63
  government action, 158–59, 162–63
  introduction, 131, 132
  lead and, 155, 174–75
  legal aspects of, 149–50
  from new sources of water, 158
  runoff and, 156–57, 162
  sewage treatment plants and, 155–56,
    157
  taste and, 150

  threats to, 155–58
  today, 150–51, 155
  watershed protection, 149, 155–59,
    162–63
  from wells, 151, 152–54, 158
Drinking water supply, 132–48
  background, 132–34
  consumption, 138–41
  drought, 140–41, 142, 148, 160
  fees for, 145
  government action, 142–48
    conservation and leak prevention,
      141–42, 144–48
    new supply, 142–44, 158
  groundwater and, 143–44
  history of, 133–34
  introduction, 131, 132
  legal aspects of, 141–42
  maps of, 136–37
  meters for, 138, 145
  "safe yield," 141
  system today, 134–38
  unaccounted for losses, 146–48
Drought, 140–41, 142, 148, 160
Dry cleaning establishments, 122
Dubos Point, 67
Dumping fees, 4, 12, 25

Earth Care Paper Company, 213
Earth's atmosphere, 89
East Hampton, Long Island, 21
East River Landing, 70
Edgemere landfills, 12
  leachate at, 7, 9–10, 197
Environmental Action Coalition, 24,
    214
Environmental Defense Fund (EDF),
    40, 157–58
Environmental impact statements, 101
Environmental Protection Agency
    (EPA), U.S.:
  air pollution and, 96, 98, 99, 105–
    106, 107, 111, 118, 121–27

asbestos and, 181, 184, 221
  drinking water and, 149–50, 162
  hazardous wastes and, 195
  lead and, 169, 175, 176
  PCBs and, 188
  solid waste and, 9, 38, 41–42
  waterways pollution and, 68, 81, 83

Fish, 50, 59, 65, 70, 81, 83, 187
Floyd Bennett Field, 67
Fluoride in drinking water, 151
Flushing Bay, 54–57
"Fly ash," 40, 41
Food and Drug Administration, U.S.,
  172
Formaldehyde, 106
Fossil fuel combustion, 111–14
Fountain Avenue landfill, 7, 9, 13, 193
Fresh Creek, 67
Fresh Creek holding tanks, 54
Fresh Kills landfill, 10–12, 41, 80,
  111
  dependence on, 3, 11
  forecast closure of, 12
  leachate at, 7, 9–10
Friends of Gateway, 215

Gasoline fumes, 108, 117
Gasoline-powered vehicles, 100–106
  alternative fuels for, 108–109
  carbon monoxide and, 101–102
  inspection programs for, 107–108
  lead and, 92, 96, 104–105, 175–76,
    177
  ozone and, 102–104
  reducing use of, 109–11
  toxics and, 105–106
Gateway National Recreation Area, 13,
  65, 73
  map of, 66
General Electric Company, 59, 186–87
General Motors Corporation, 108
Gerrard, Michael, 77

Giardia lamblia, 148
Global warming, 8, 74, 124–25, 139
Goldin, Harrison, 21
Gowanus Canal, 54, 58
Graham Beach, 49
Gravesend incinerator, 32, 33
Great Kills Beach, 14
Greenhouse effect, 8, 74, 124–25,
  139
Greenpoint incinerator, 32, 33
Groundwater pollution, 8

Hazardous waste dumps, 10, 12, 13,
  18
Hazardous wastes, 189–97
  background, 189, 192
  disposal of, 193–95
    inactive sites for, 194
  generation of, 192
  government action, 196–97
  household, 190–91
  legal aspects of, 195–96
Health Department, New York City,
  196, 209, 217, 222
  drinking water and, 152, 153, 162
  landfills and, 17–18
  lead and, 170, 174–75
  water pollution and, 49
Health Department, N.Y. State, 92,
  126, 215, 217
  drinking water and, 142, 150, 162,
    163
  medical waste and, 14, 16
Hepatitis, 15
High-sulfur fuel, 112–14
Highway Users Federation, 108
Hillview reservoir, 135, 138
Hinchey, Maurice, 30
Hoffman, Abbie, 125
Hospital incinerators, 16–17, 37, 38, 88,
  116, 177
  list of, 115
Household toxics, 190–91

Hudson River, 49, 70
  drinking water from, 132, 138, 139, 142–43, 158, 159, 160
  PCBs in, 59, 186–87, 189
Hudson River Center, 70
Hydrocarbons, 100, 103, 104, 106, 108, 120
Hydrogen chloride, 37, 38

Idlewild landfill, 18–19
Incineration, 31–43, 116
  apartment house, 32–33, 37–38, 43, 116, 177
  background, 32
  crisis concerning, 3–4
  as divisive issue, 31
  government action, 42–43, 142
  health and environmental risks, 4, 36–41
    air pollution, see Air pollution, from incinerators
    ash disposal, 12, 33, 34, 40–41
    lead and, 177–78
    resource recovery incinerators and, 38–40
  hospital, 16–17, 37, 38, 88, 115–16, 177
  introduction, 31
  legal aspects of, 41–42
  proposed, 3–4
  today, 32–33
  tomorrow, 33–36
Indirect dischargers, 61–62
  Clean Water Act and, 63–64
  government action on, 64, 68
Indoor pollutants, 123–27
  common sources of, 123
Indoor Radon Abatement Act of 1988, 127
INFORM, 109
Inside Story: A Guide to Indoor Air Quality, 218

"Intensive recycling," 24
Interstate Sanitation Commission, 51, 110

Jacob Riis Park, 73–74
Jamaica Bay, 49
  development of, 73
  landfills along, 13, 16
  leachate in, 7–8, 65, 193
  toxic water pollutants in, 65–67, 197
Jamaica sewage treatment plant, 51
Jamaica Water Supply Company, 138, 144, 151, 152–54, 158
Jerome Park reservoir, 135
Jorling, Thomas C., 34, 108

Kass, Stephen L., 77
Kennedy, Robert F., Jr., 156
Kennedy Airport, 65, 67, 120–21
Kensico reservoir, 135
Kiley, Robert, 109
Koop, C. Everett, 91

Labor Department, N.Y. State, 203, 221, 222
Labor Department, U.S., 203, 209
La Guardia Airport, 120–21
Landfills, 5–19
  background, 5–7
  Brookfield Avenue, 17–18
  Edgemere, 12
  environmental and public health aspects, 7–9
    air pollution, 8–9
    groundwater pollution, 8
    surface water pollution, 7–8
  Fountain Avenue, 13
  Fresh Kills, 9–12
  health risks posed by, 4
  legal aspects of, 9
  medical waste, 14–17
  other closed, 18–19

Pelham Bay, 18
Pennsylvania Avenue, 13, 16
report card on, 9–19
sites of, 6
see also specific landfills
Leachate, 7–8, 10, 13, 65
Lead, 168–78
  in air, 104–105, 175–78
  in ancient Rome, 169
  background, 168–69
  exposure routes, the law, and
      government action, 170–78
  in gasoline, 92, 96, 104–105, 175–76,
      177
  health effects of, 169–70
  lead belt, 173
  in paint, 171–73
  poor people and, 168, 171, 172
  in solid waste, 7, 8, 18, 29, 36, 37,
      40, 41
  sources of, 169
  in water, 60, 63, 155, 174–75
Leffler, Sheldon, 21, 43
Lindane, 126
Little Neck Bay Beach, 49
Locust Point Beach, 49
Low-sulfur fuel, 88, 112–14

Mahopac sewage works, 156, 157
Marine Protection, Research and
      Sanctuaries Act of 1972, 84
Markets for recyclable items, 27
Maspeth incinerators, 32, 33
Mayor's Intergovernmental Task Force
      on New York City Water Supply
      Needs, 142, 147–48
Medical wastes, 4–5, 14–17, 80
  incineration of, 16–17, 37, 38, 88,
      115–16
Mercury, 105
  as occupational toxin, 203, 209
  in solid waste, 7, 36, 37

  in water, 59, 65
Messinger, Ruth, 21
Methane, 8
Mexico City, 87
Microwave News, 201
Midland Beach, 14, 49
Mill Basin, 73, 77
Miller Field Beach, 14
Molinari, Guy, 111
Montefiore Hospital, 222
Motor Vehicle Air Pollution Control
      Act, 106
Motor vehicles, *see* Air pollution, from
      motor vehicles
Mount Sinai Medical Center, 197, 210,
      223
Municipal Arts Society, 72, 216
Murray Hill reservoir, 133
Muscoot reservoir, 156
Muskie, Edmund, 96

National Academy of Sciences, 98
National Institute for Occupational
      Safety and Health, 179
National Oceanic and Atmospheric
      Administration, 60, 75
National Park Service, 13, 73–74
National Resources Defense Council
      (NRDC), 215, 216, 218, 222
National Testing Laboratories, 220
Newark airport, 121
New Croton reservoir, 135
New Jersey vs. *City of New York*, 80
Newtown Creek, 54, 193
Newtown Creek sewage treatment
      plant, 51, 52, 53, 62
New York Academy of Medicine, 163
New York Bight, 83
New York City Street Tree Consortium,
      217
New York Committee on Occupational
      Safety and Health, 210

New Yorker, 104
New York Lung Association, 217
New York Newsday, 159
Nickel:
    in solid waste, 7, 29
    in water, 60, 62, 65
Nitrogen dioxide, 37, 99, 112, 114
Nitrogen oxides, 100, 103, 105, 113,
    120
Non-Ferrous Processing Corporation,
    176
"Nonpoint source" pollution, 156–58
North River Water Pollution Control
    Plant, 51, 61, 120

Occupational Safety and Health Act
    (OSHA) of 1970, 181, 203, 209
Occupational Safety and Health
    Administration, 126, 209, 222
Occupational toxics, 197–210
    background, 197–98
    exposures, 198–203
        table of, 204–208
    government action, 203, 209–10
    legal aspects of, 203
    prevention and, 203
Ocean dumping, 32, 80–84
Ocean Dumping Act, 84
Office paper recycling, 26
Oppenheimer, Dr. John, 110
Organic Gardening, 214
Owl's Head sewage treatment plant, 51,
    52
Ozone, 102–104, 105
Ozone smog, 8, 88, 93, 103–104, 108,
    117–18

Paerdegat Basin, 58, 67, 73
Paint, lead in, 171–73
Parks Department, New York City, 67,
    215, 216
Particulate pollution, 36–37

from commercial, industrial, and
    residential sources, 112, 113
from diesel engines, 88, 93, 96–100,
    105, 107
Pelham Bay landfill, 18
Pennsylvania Avenue landfill, 7, 13, 16,
    193, 196–97
People's Environmental Program, 27
Perchloroethylene, 122
Petroleum marketing, 117
Planning Department, New York City,
    76, 77
Plastics, recycling of, 23, 28–29
Pliny, 169
Plumbing and lead, 155, 174–75
Plumbing Foundation, 146, 175
Plutonium scare, 160
Polychlorinated biphenyls (PCBs), 185–
    89
    background, 185–86
    chemicals in your community, 188
    in electrical equipment, 187
    exposures in New York, 186–87
    health and environmental effects, 186
    in the Hudson, 59, 186–87, 189
    legal aspects and government action,
        187–88
    in solid waste, 8, 13, 18
    in water, 59, 60, 63, 65, 83, 158, 193
Polyvinyl chlorides (PVCs), 17
Port Authority of New York and New
    Jersey, 120–21
Port Liberte Partners, 81
Power plants, 111–14
Procurement and recycling, 27, 30
Public Development Corporation
    (PDC), 73, 77
    project wish list of, 78–79
Public Health Service, U.S., 170
Putnam Hospital, 156
Pymm Thermometer Corporation, 203,
    209

Quanta Resources Corporation, 193
Queensboro Lung Association, 217

Radiac Research Corporation, 199–201
Radon, 125–26, 127
*Radon: A Homeowner's Guide to
    Detection and Control*, 218
Raritan Bay, 60
Ravenswood power plant, 114
Reagan, Ronald, 51, 176
Recycling, 19–30
    current status of, 21–30
        bottle bill, 23
        commercial, 24–26
        markets and, 27
        office paper, 26
        plastics and, 23, 28–29
        procurement and, 27, 30
        recycling centers, 26–27
        residential, 24
        table of programs, 22
    introduction, 19
    legal aspects of, 19–21
    as percent of waste stream, 5
Recycling centers, 26–27
Red Hook sewage treatment plant, 51,
    61
Regional Plan Association, 155
Residential recycling, 24
Resource Conservation and Recovery
    Act (RCRA) of 1976, 9, 195
"Resource recovery" incinerators,
    38–40
Rising seas, 74
Rivers and Harbors Appropriations Act
    of 1899, 75
Riverwalk, 70
Rockaway Peninsula, 12
Rogers, Paul, 96
Rome, ancient, 169
Roosevelt, Theodore, 211
R2B2, 26–27

Safe Drinking Water Act of 1974, 149
Safety Kleen Corporation, 201
Sandler, Ross, 101
Sanitation Department, New York City,
    18, 43, 97, 193
    incinerators and, 35–36, 116
    recycling and, 19–21, 24–26, 27, 212,
        214
San José, California, 146
Scenic Hudson, 143
Schoenbrod, David, 101
Seagate Beach, 49
Sewage pollution, 48–58
    background, 48–49
    combined sewer overflow, 53–55, 58
    in drinking water, 155–56, 157
    in Flushing Bay, 56–57
    government action, 50–53
    holding tanks for, 54–55, 56, 58
    impacts of, 49–50
    in Jamaica Bay, 65
    legal aspects of, 50
    treatment plants, 50–53
Sewage sludge, 80–81, 83
Sewage treatment plants, 48–49
    air pollution from, 118–20
    drinking water pollution from, 155–
        56, 157
    government action and, 50–53
Sheepshead Bay, 73
Shellfish, 49–50, 59, 65, 81, 83
Sludge, 80–81, 83
Solid waste, 1–43
    epilogue, 212–14
    incineration, *see* Incineration
    landfills, *see* Landfills
    New York State's plan for, 20
    overview of, 405
    recycling, *see* Recycling
    statistics on, 4–5
    as three-part crisis, 3–4
        capacity crunch, 3

Solid waste (cont'd)
   environmental and health risks
      from landfills and incineration, 4
   logistics, 3–4
   waste reduction, 30
   where it goes, 5
South Beach, 14, 49
South Shore landfill, 18–19
Space heating, 111–14
Spring Creek, 67
Spring Creek holding tanks, 54
Staten Island, 74
   air pollution in, 93, 110–11
State Pollutant Discharge Elimination
   System (SPDES), 68
Stein, Andrew, 217, 222
Stewart airport, 121
Suburban Water Testing Laboratories,
   220
Sulfates in drinking water, 157–58
Sulfur dioxide, 37, 88, 99, 105, 112–14
Sulfur oxides, 38
Superfund, 193, 195, 196
Superfund Amendments and
   Reauthorization Act (SARA) of
   1986, 195
Supreme Court, U.S., 80, 102, 133, 141
Surface water pollution, 7–8
Swimming, banning of, 14–16, 49

Times Beach, Missouri, 37
Tobacco smoking, 91–93, 179
Toilets, water-conserving, 141–42, 145,
   147
Toluene, 63, 121, 161
Toxic air pollution, 105–106, 122
Toxic drinking water pollution, 148, 152
Toxics, 165–210
   asbestos, see Asbestos
   epilogue, 221–23
   hazardous wastes, see Hazardous
      wastes

   household, 190–91
   introduction, 167–68
   lead, see Lead
   media and, 167
   occupational, see Occupational toxics
   PCBs, see Polychlorinated biphenyls
      (PCBs)
   threats from, 168
Toxic Substances Control Act of 1976,
   181, 188
Toxic waste disposal, 12
   ash residue and, 40–41
Toxic waterways pollution, 58–68
   background, 58–59
   government action, 64, 68
   impacts of, 59–60
   in Jamaica Bay, 65–67
   legal aspects of, 63–64
   types of polluters, 60–63
      direct dischargers, 60–61
      indirect dischargers, 61–62
      urban runoff, 62–63
Transit Authority, New York City, 109,
   192
Transportation Alternatives, 218
Transportation Department, New York
   City, 110
Trap oxidizer, 109
Triboro Coach Corporation, 108
Trichloroethylene, 8, 152
Trihalomethanes in drinking water, 151,
   152, 161
Trump, Donald, 72
Trump City, 69–70
   impact of, 71–72
Trust for Public Land, 73
2,3,7,8-TCDD, 37
Typhoid, 148

Ulano Corporation, 121–22
Urban Air Toxics Assessment Project,
   111

Urban runoff, 62–63
  Clean Water Act and, 64
Urban visibility, 99–100

Video display terminals (VDTs), 201,
    203
Village Green Recycling Team, 27
Vinyl chloride, 126

Wards Island sewage treatment plant,
    51, 53
Waste reduction, 30
Water, drinking, see Drinking water
    quality; Drinking water supply
Water Board, New York City, 145
Waterfront Revitalization Program, 76,
    77
Water Pollution Control Act of 1972, see
    Clean Water Act of 1972
Water Resources Planning Council, 140
Watershed protection, 149, 155–59,
    162–63
Water Supply Act of 1905, 141

Water Supply tunnel number three,
    138, 146
WaterTest, 220
Waterways, 45–84
  the coast, see Coast, the
  epilogue, 215–17
  introduction, 47–48
  ocean dumping, 80–83
  sewage pollution, see Sewage
    pollution
  toxic water pollution, see Toxic
    waterways pollution
WE CAN, 213
Weiss, Ted, 112
West Harlem Environmental Action,
    120
Westpride, 72
White Lung Association, 184
Williams, Henry, 114
Williamsburg, 199–201

Zinc, 60, 62, 63, 65
Zoning Resolution, 76

ABOUT THE AUTHORS

Eric A. Goldstein is one of New York City's leading environmental lawyers. A senior attorney with the Natural Resources Defense Council, he has been involved for more than a decade in many of the city's environmental protection battles, most notably as a lobbyist and litigator on air pollution from motor vehicles, industrial facilities, and garbage-burning incinerators. He gained national attention in the early 1980s for spearheading the nationwide campaign to reduce levels of toxic lead in gasoline. Mr. Goldstein, who holds degrees from Hofstra University Law School and New York University School of Law, serves on the boards of the New York Lung Association, the New York City Environmental Control Board, the New York State Solid Waste Management Board, and the Environmental Planning Lobby. Born and raised in New York City, Mr. Goldstein resides on Manhattan's West Side.

Mark A. Izeman was a senior researcher at the Natural Resources Defense Council from 1986 until enrolling in New York University Law School as a Root-Tilden-Snow public interest scholar in the fall of 1989. During that period he was one of the principal citizen lobbyists pressing for enactment of New York City's landmark 1989 mandatory recycling law. Despite his busy law school schedule, Mr. Izeman has continued to work at NRDC as a consultant on solid waste and water quality issues affecting New York. He received his undergraduate degree from Brown University and now lives in Brooklyn.

ALSO AVAILABLE FROM ISLAND PRESS

Ancient Forests of the Pacific Northwest
By Elliott A. Norse

The Challenge of Global Warming
Edited by Dean Edwin Abrahamson

The Complete Guide to Environmental Careers
The CEIP Fund

Creating Successful Communities: A Guidebook for Growth Management Strategies
By Michael A. Mantell, Stephen F. Harper, and Luther Propst

Crossroads: Environmental Priorities for the Future
Edited by Peter Borrelli

Environmental Agenda for the Future
Edited by Robert Cahn

Environmental Restoration: Science and Strategies for Restoring the Earth
Edited by John J. Berger

The Forest and the Trees: A Guide to Excellent Forestry
By Gordon Robinson

Forests and Forestry in China: Changing Patterns of Resource Development
By S.D. Richardson

From *The Land*
Edited and compiled by Nancy P. Pittman

Hazardous Waste Management: Reducing the Risk
By Benjamin A. Goldman, James A. Hulme, and Cameron Johnson
for Council on Economic Priorities

Land and Resource Planning in the National Forests
By Charles F. Wilkinson and H. Michael Anderson

Last Stand of the Red Spruce
By Robert A. Mello

Natural Resources for the 21st Century
Edited by R. Neil Sampson and Dwight Hair

Overtapped Oasis: Reform or Revolution for Western Water
By Marc Reisner and Sarah Bates

The Poisoned Well: New Strategies for Groundwater Protection
Edited by Eric Jorgensen

Race to Save the Tropics: Ecology and Economics for a Sustainable Future
Edited by Robert Goodland

Reforming The Forest Service
By Randal O'Toole

Reopening the Western Frontier
From *High Country News*

Research Priorities for Conservation Biology
Edited by Michael E. Soulé and Kathryn Kohm

Resource Guide for Creating Successful Communities
By Michael A. Mantell, Stephen F. Harper, and Luther Propst

Rivers at Risk: The Concerned Citizen's Guide to Hydropower
By John D. Echeverria, Pope Barrow, and Richard Roos-Collins

Rush to Burn: Solving America's Garbage Crisis?
From *Newsday*

Saving the Tropical Forests
By Judith Gradwohl and Russell Greenberg

Shading Our Cities: A Resource Guide for Urban and Community Forests
Edited by Gary Moll and Sara Ebenreck

War on Waste: Can America Win Its Battle with Garbage?
By Louis Blumberg and Robert Gottlieb

Western Water Made Simple
From *High Country News*

Wildlife of the Florida Keys: A Natural History
By James D. Lazell, Jr.

For a complete catalog of Island Press publications, please call our toll-free number 1-800-828-1302 or write to Island Press, Box 7, Covelo, CA 95428.

Island Press Board of Directors

PETER R. STEIN, CHAIR
Senior Vice President
The Trust for Public Land

DRUMMOND PIKE, SECRETARY
Executive Director
The Tides Foundation

WALTER SEDGWICK, TREASURER

ROBERT E. BAENSCH
Director of Publishing
American Institute of Physics

PETER R. BORRELLI
Editor, *The Amicus Journal*
Natural Resources Defense Council

CATHERINE M. CONOVER

GEORGE T. FRAMPTON, JR.
President
The Wilderness Society

PAIGE K. MACDONALD
Executive Vice President/
Chief Operating Officer
World Wildlife Fund/The Conservation Foundation

CHARLES C. SAVITT
President
Island Press

SUSAN E. SECHLER
Director
Rural Economic Policy Program
Aspen Institute for Humanistic Studies

RICHARD TRUDELL
Executive Director
American Indian Lawyer Training Program